Technology and Social Change

Technology and Social Change

Edited with an Introduction by
Wilbert E. Moore

A NEW YORK TIMES BOOK

Quadrangle Books
CHICAGO

301.24308
m825t

Library of Congress Catalog Card Number: 76–156336
SBN Cloth 8129–0214–9
SBN Paper 8129–6178–1

The publishers are grateful to the contributors herein
for permission to reprint their articles.

Contents

31395

vii • *Contents*

Technology and Social Change

Introduction

MACHINES AND TOOLS fascinate Americans. This does not make them different from Englishmen or Germans or Japanese or Russians or others who have grown up in a technological era. Nor is the fascination always a simple matter of pure affection, of liking the capacity to do things that manual labor will not do. The mastery of the machine appeals to the housewife with her nominally automatic home appliances, to the home craftsman with his woodworking or metal-cutting power tools, and to all of those automobile drivers who use their vehicles as extensions of their personality or, not infrequently, as weapons. But when the machines do not work, when complicated mechanisms beyond the knowledge of the ordinary user fail to function, the fascination quickly gets translated into exasperation. The machine becomes an enemy, with all the aesthetic appeal of a basket of poisonous snakes.

Man's love-hate relationship to the machine is pervasive in the modern world. It does not diminish with the growing complexity of mechanical processes. If workers once smashed the textile machines which they feared—sometimes correctly—were displacing them from jobs or handicraft markets, the contemporary citizen voices his complaints about the errors (or perhaps the uncanny accuracy) of computers, and shares with friends some favorite scheme for outwitting the infernal machines. Two opposed camps have come into being: the planners and dreamers who plot or predict ever more wondrous mechanical marvels, and those who vainly pine for a simpler relation of man to his fellows and to his environment. Yet the naysayers and the Jeremiahs generally appear to be losing, and the machines and their advocates to be winning.

The machines do not really win, of course; only the inventors, investors, and other beneficiaries of technical change do. The assignment of good or evil, of malice or good will to the machine may be an acceptable form of poetic license, but it will not do as a basis for reasoned analysis. The near-human (sometimes superhuman) capacities of the electronic computer have added a new verisimilitude to an old image of the machine personified, but in attributing a moral quality to machines we are still playing with a metaphor that finally does not hold up.

Man is still in charge of his destiny, though some men are more in charge than others, and part of man's destiny turns out to be the unwanted, even dangerous by-product or unforeseen consequences of the very attempt to control his environment. The enemies of technology are no longer limited to those whose livelihood has been immediately threatened, or to the unbelievers who thought that the automobile was a passing fad and that planes would never get off the ground. The costs as well as the benefits of technology are under increasing scrutiny and a cause for growing concern.

The Meaning of Technology

The first step in any sober appraisal of technology in modern society and its position with respect to social change is to clarify what is meant by technology itself. In the preceding paragraphs we have come very close to a common error in discussions of technology, and that is the equation of technology with the machine. The machine is only the most conspicuous and tangible manifestation of man's ingenuity in adapting to his environment and altering it to suit his own purposes. But the machine is not necessarily the most important, or even the most dangerous, result of "tinkering with nature."

Technology is best understood as the application of knowledge to the achievement of particular goals or to the solution of particular problems. The knowledge to be applied—and how to apply it successfully is also a part of the store of knowledge—may involve abstract principles and scientific "laws," or be merely experimental, a bit of proven lore. It must, however, be observationally based and at least approximately verified. Technology differs from religious acts and beliefs in that the goals or purposes it aspires to achieve, as well as the means for attaining them, are this-worldly and subject to observational verification. It differs from magic in that its means are also "rational" and instrumental, in the ordinary sense of accounting for causation, and do not depend on supernatural or mystical sources. The difference may be a subtle one, of course, as many a beneficiary or victim of technical processes may not in fact understand them, with the consequence that they become, for him, magical. If I press this button, here, the computations are printed out there. Yet many people do understand the process in very matter-of-fact terms, and without reference to vague and mysterious forces.

Technology may also be distinguished from science, though with greater difficulty. "Pure" science is presumably concerned only with the discovery of truth—the truth that is subject to observational and objective verification—and with the ordering and generalizing of mere facts into principles and laws of relationship and sequence. Thus, while scientists are nominally not concerned with utility, technologists have almost no other excuse for their activities. The distinction is a valid one in the abstract, and indeed often in the concrete. Scientific principles may be valid without being useful—witness much of astronomy or paleontology—and the inventor or imitator of procedures for starting fires or forging bronze or iron need not know the scientific principles that explain the rationale for techniques that work.

It remains true, however, that much of science turns out to

be practical, or have practical applications, whatever the intentions of those who formulate scientific principles and relate them to other, equally scientific, principles. And it is also true that some principles are discovered and others amended in practice, even if testing the principles was not the aim of the technologist. Technology may be too elegant a word for the practical knowledge of the individual inventor, tinkerer, or mechanic. Perhaps its adoption was inevitable, for in the modern world even useful knowledge tends to become formalized and generalized. Significantly, many former schools of engineering now style themselves schools of "engineering sciences."

In an evolutionary sense, it may be accurate to characterize man as "the tool-using animal," for tools and weapons may well have been the first manifestations of human ingenuity in coping with a hostile or at least a challenging environment. But fire is also of ancient and unknown origin, and it scarcely qualifies as a primitive machine (though in one form or another it was, of course, later used to power true machines). If such glib characterizations of man are helpful at all, he is better designated as the problem-solving animal. The knowledge potentially available for application is by no means limited to mechanical relations, even if those become very sophisticated, as they are in heat transfers, electronic circuitry, or nuclear fission and fusion.

A further point of clarification is appropriate here. The equation of technology with its mechanical manifestations is not merely too limited a conception of useful knowledge; it is also a wrong one. As knowledge of art or craft, which the word "technology" literally means, it exists in men's minds or in symbols and drawings on paper well before it takes form in processes or products or other palpable results. Technology is no more "material" than beliefs or attitudes or aesthetic responses to perceived beauty or ugliness; it is simply oriented toward "practical" affairs. Thus the distinction drawn by William F. Ogburn, a distinction famous at least among sociologists, between "material" culture and "nonmaterial" or "adaptive" culture, rests upon a fundamental error of conception. (Ogburn never truly repented his error, as the careful reader of his essay in this book will see—the essay was written more than twenty years after his original formulation—but he did not let it stand in the way of otherwise cogent thinking and investigation.) Useful knowledge may be manifested in a tool, a medical cure, or the successful pleading of a case in court. Even the machine need not move mountains or grind out gadgets. It may be used to destroy a brain tumor or electronically produce noises that pass for music. It is perhaps well to keep this elementary and essentially semantic point in mind, lest we be numbed by the mechanical metaphor. Technology may be, and usually is, shared, and thus to some degree is collectivized and depersonalized, but it is neither a disembodied nor a nonhuman force. It is the product of human ingenuity, and without human purpose behind it, it is not the cause of anything.

The varieties of technology depend on what is identified as a *problem,* that is, a situation for which remedies or solutions are at hand or may be seriously attempted—rather than as a *given,* which is a condition defying human intervention. The technical state of mind is secular. It has scant patience with Fate, or with Divine Will, or even with tradition, the wise teachings of the founding fathers, or the shared but unexamined wisdom which "everyone knows."

Technology has a history. And in the properly full sense of the term "technology," its history provides some basis for a sweeping evolutionary view of the accumulation, extension, and diversification of useful knowledge. In broad outlines the progression would run from mechanical manipulation to chemical and metallurgical skills, thence to biological intervention (for example, in animal husbandry and sophisticated agriculture, but proceeding to medical arts), and, finally, to social technology, ranging from applied psychology to national economic planning. But this grand view of the extension of attempts to be informed and to be "practical" is wrong in details, and those details are by no means trivial. For example, the domestication of animals was prehistoric, as were fire, tools, and pots; so, indeed, was symbolic language, which may be the most important social invention of mankind. Yet it does appear that in terms of time and effort, and of results, inanimate nature was more amenable to manipulation, more vulnerable, if you will, than animal nature, and both were more subject to calculating rationality than the forms and processes of social interaction and organization.

If this interpretation holds as a broad generalization, it does not by any means lead to the conclusion that technology leads all forms of human endeavor, and other social interests and practices follow and adapt themselves to technological change with various degrees of tardiness; but only that some kinds of technology may lead others. Although I shall address myself to the central problem of the relation of technology to social change in the concluding section of this essay, it is worth noting here that the transformation of givens, or conditions inherent in nature, into problems is first of all a perceptual and attitudinal matter. Essentially accidental innovations may occur and find acceptance, but the devotion of time, energy, and other resources to deliberate change requires a prior commitment. That commitment is a social decision, not a technical one.

The importance of technology to human existence and comfort is not degraded by observing that technology is not an inanimate or disembodied force that operates independently of human purpose. That so many otherwise sensible historians and social analysts, among them Karl Marx and Thorstein Veblen, should have slipped into the error of granting technology primacy among human activities is not totally surprising, nor was their error total. First, technical knowledge is cumulative, with rare exceptions (such as the mostly exaggerated "lost arts"). Thus, technical change may seem to "feed on itself," each addition to the store of information and skills multiplying the possible new combinations. This principle of technological accumulation does not assure continued change, but does provide favorable conditions for it. Second, many technical problems are not finally solved, and indeed many technological solutions themselves create new problems—a subject to which I shall return. Neither the inanimate nor the animate environment is truly under man's control— witness the still-present rampages of weather and disease, not to mention the growing problem of "using up" the environment through excessive population growth. This principle of persistent environmental challenge further helps account for the notion that technology leads other kinds of human endeavor, as, at times, it indeed does.

Man's Relation to "Nature"

All animal species must adapt to environmental conditions if they are to survive, and selective adaptation is a key variable in evolutionary theory. Adaptation, moreover, need not be entirely passive and physiological. The environment may be used (and parts of it used up, such as in the defoliation of plants by insects and browsing mammals), and thus be altered in temporary or more enduring ways—for example, the creation or destruction of birds' nests, ant hills, or beaver dams as multi-generational dwelling places. How much of this nonhuman behavior is instinctive and how much passed along by some form of learning is a problem that need not seriously detain us, for the central point is that the human species is not alone in its capacity for "manipulation" of the environment. Yet it is the long success and rapidly growing power of man's capacity for manipulation that mainly accounts for the conspicuous alteration of the original physical environment, the construction of a new material environment that is increasingly man-made, and, not least, the successful competition for space and other resources to the point where many other animal species have either become extinct or rendered virtually doomed to make way for man's needs.

Since man's store of technical knowledge and skills *is* accumulative, and thus greatly different in degree if not in kind from that displayed by other animal species, it is not surprising that technology is often viewed primarily in terms of the mastery of the physical environment. That mastery, we have noted, is not complete, but it is impressive nonetheless. Each year even further alterations become technically possible—for example, the maintenance of all-year sea shipping routes in far northern latitudes, the irrigation of otherwise infertile deserts with freshened sea water, or the creation of cities on the open sea by constructing artificial barrier reefs and lagoons. The only restraint on such innovations is the small one of cost.

Man's success in the mastery of nature is not only incomplete, it is unequal. Generally, this mastery has been greater over visible and palpable phenomena than over such minute entities as molecules, atoms, and subatomic particles or microbes and viruses. This generalization would hold even more firmly if we added the qualification "accessible" phenomena. The sea remains little explored or exploited, and, despite unmanned flights around nearby planets and manned landings on the moon, so does outer space. Again, cost is a considerable restraint on known or not wildly improbable techniques. In the case of space travel there is a further problem: distance, which has serious implications for time. The speed of light (the velocity at which matter becomes energy) appears to be a physical absolute. Given man's life span, a round trip to a star two thousand light years away has little appeal; other forms of suicide are easier and cheaper.

From these giddy heights and depressing depths, let us return to more terrestrial matters. The relation of the human species to its nonhuman environment has always been interactive in some form and degree. The livestock of nomadic herdsmen exhaust a pasture land and move along with their owners to greener pastures. Cultivators in tropical rain forests slash the trees and bushes, burn them, plant in the ashes, but then move to a new site as rains leach the soil of necessary plant nutrients. These archaic (though still extant) forms of exploitation of natural resources are restricted by man's own excessive reproductive capacity and competing territorial claims on space that grows ever more scarce.

The direct interaction between human beings and "nature" is reduced almost to the vanishing point for modern urban dwellers. In cities, the physical environment is almost completely man-made: houses, apartments, streets, means of transportation, places of work, and places of play. Parks, streams, harbors, and beaches are either man-made or man-polluted, or both. Weather remains partly intractable, but heating, air conditioning, foul-weather clothing, and chemical snow removal soften the impact. If the urban dweller notices the sky at all, he is likely to see it only through a smog darkly.

The physical composition of the earth remains relatively fixed, though long-term climatic changes and geological changes of both short and long term affect its usability or accessibility. Current technology and the consequences of past human actions have a far greater effect on resources, however. Some of those consequences are essentially irreversible. Soil washed into the ocean, whether from natural or man-made erosion, is essentially irretrievable; the residue from burning fossil fuels (coal, natural gas, and petroleum) is worthless or worse. Metals are also a "wasting" rather than a replaceable resource, though they are generally not literally destroyed through use. (The metal mines of the future may be the junk yards and land fills of today.) The rate of consumption of relatively fixed supplies of raw materials now far outpaces the rate of population growth, and that growth is itself a part of the problem, since space is also an increasingly scarce resource.

On the subject of exhaustion of resources, the pessimists seem to be currently in the ascendancy. (I say "seem to be"; it may be that they are simply noisier than the optimists.) The pessimists clearly have the edge in any argument that attends to population growth as an important consideration, for the extrapolation of present growth rates for only a few hundred years would produce a situation with no more stacking room at tolerable altitudes for breathing, standing room by that time having ceased to be an attainable luxury. Such a situation will not come to pass, but it is an instructively insane vision, for it points to the insanity of denying that a problem exists.

In coming to grips with the problem of exhaustion of resources, the optimists point to substitution of materials, including synthetic —that is, man-made—ones, along with unexploited resources deep in the earth or deep in the sea. Which brings us to an important point. The very identification of what *is* a resource, and therefore its potential quantity, quality, and use, depends on the state of the useful arts—that is, on technology itself. Cost, of course, is always a factor, but that is in large part a question of the allocation of resources, now using the term in the extended sense of how much time, energy, knowledge, skills, usable raw materials, and instruments will be devoted to the conversion of any resource into products. Oil shales are either unproductive rocks or a potential source of petroleum; uranium ores are either a source of a coloring agent for special pottery or a valuable raw material for atomic power; bauxite is either an especially infertile type of clay or the source of aluminum; air is either a "free good" or a grand public dump for airborne solids and noxious fumes. New knowledge will undoubtedly continue to produce new resources as well as increase the efficiency with which known resources are exploited or destroyed.

If the struggle with nature has been partially won, the scars and other costs of battle grow apace. The fairly recent concern for ecology—the relation of organisms to their environment— has become a new rallying point for critics of social arrangements and social policy. Ecology was once a rather limited specialty in the fields of biology and sociology, but now the term and its underlying notions of interplay and change have become more public, and more controversial. Critics point to the individual and social costs of the man-made environment—costs often not borne by the engineers, the manufacturers, or even, in adequate measure, the users and consumers.

The benefits of technology can scarcely be denied: increased comfort and affluence, better health and longer life, and, not least, leisure to develop new knowledge. It is not sensible, and fortunately scarcely feasible, to return now to the simple life. Life in a true state of nature, as Thomas Hobbes noted over three centuries ago, is "solitary, poor, nasty, brutish, and short." The artificiality of the contemporary environment does not make it automatically superior. To the pompously inane admonition, "You can't fight progress," there is always the suitably querulous reply, "Who says it's progress?"

Productive Modes and Relationships

Some of the most bitter social criticism of technology has centered on the servitude of man to the machine. Certainly the so-called Industrial Revolution was primarily based on the substitution of inanimate sources of power for human muscle and dexterity. This substitution entailed the movement of nonagricultural production out of the household and into the factory. Factory production, in turn, entailed subjection to the discipline of bosses or their hirelings. A significant part of that discipline was temporal: starting times, stopping times, and the pace of activities in between. The clock is the primordial machine of an industrial civilization.

The course of the mechanization of production is instructive. Not only was much of manual labor taken over by machines, but the worker was also paced by the machine. The assembly line perhaps marked the high point in the mechanization of man. Almost every bit of discretion, judgment, or skill in any significant sense was removed from the job assignment. The workman was left with no choice in the accomplishment of his tasks.

This appears to be a clear case of a major social change being dictated by an advancing social technology. The worker was separated from his family and neighbors in time and place, treated impersonally at his place of work, and indeed generally used as a kind of inefficient machine. But a major social precedent for such arrangements should not escape us: behind everything was capital concentration beyond the ordinary worker's capacity to afford. In the classic formulation of Karl Marx, the worker was "expropriated" from the means and instruments of production and owned only his own labor. Although Marx was more than slightly confused on this point, this was an arrangement in the control of property, and not itself technologically determined. The pooling of labor and capital on a large scale depended on mass production and mass consumption. And these, in turn, depended on the prior social invention of the monetary market. Without a market, the machine remains useless—it is no more than a rather bizarre sculpture.

Even the minute specialization of tasks, often criticized as removing all vestiges of workmanship, was not solely the result of machine technology. It has often happened that organizational planners and managers have pushed task specialization to such a point of short-cycle repetitive routine that finally the engineers could see a machine doing a better and more reliable job. Automation appears to be merely the next step in the long process of substituting mechanical processes for human ones. Its implications for the world of work are quite different, however. Automation involves displacement of labor, sometimes to a substantial degree. This entails social costs of no small magnitude. But the displacement is by no means total. The workers in a highly automated plant are no longer so visibly and cruelly the servants of the machine; they become, in a limited sense, its masters. Workers manipulate the productive process by buttons and powered levers; they monitor control boards; they diagnose ills and apply therapy; they invent, install, "de-bug," and supervise.

In terms of skill distributions in the labor force, the long-term effect of combining an ever-growing social and physical technology in the production, packaging, and even the distribution of an incredibly varied array of goods represents an *upgrading* of labor's role. Together with some proportional displacement from all types of physical production (agriculture, mining, and manufacturing), the predominant character of labor has changed. Most work is no longer physically heavy, dirty, or sweaty, though many jobs are still hazardous in both obvious and subtle ways. The machine has not made man obsolete; it has made automatons obsolete.

We have steadily moved toward a "service economy," not only in the "products" that reach the consumer but also in the kinds of jobs actually performed in organizations that appear to be devoted to manufacturing. The most rapidly growing proportion of employees in industrial corporations is comprised of scientists, technologists, and those quasi-technologists who profess to have special competence in legal, administrative, consumer, and public relations work. In terms of both manpower and economic support, the "knowledge industry" is growing more rapidly than virtually any other part of the American economy (with the possible exception of military expenditures, which have a very high technological component). Thus, metaphorically speaking, technology feeds on itself and grows fat.

The result of all this is a tremendous multiplication of human effort, accompanied by an actually reduced subservience to the machine. A principal cost, however, has been increased subservience to a social machine: the administrative organization or bureaucracy required to keep it all humming. Order, discipline, and hierarchy are the watchwords of complex organizations, for without these restraining forces the coordination of specialists would be difficult if not impossible. Bureaucracy represents an elaborately developed form of social technology. Its forms and procedures may be modified here and there to encourage initiative and creativity. Its mission may be broadened to include "development of human potential" or the reluctant recognition of various social responsibilities. Yet most of us are governed by administrators, and they in turn are presumably governed by rules.

The general upgrading of the skill levels of the labor force, including rising educational requirements for employment, means a sharp decline in the demand for relatively unskilled workers. Some workers are literally displaced by changes in product and process and cannot find new employment. Young aspirants for an ordinary place in the world of work may not find jobs. The skills offered by some workers are simply not in demand; other potential workers are unskilled and may find only casual employment or none. Educational requirements may be arbitrary, but they are nonetheless real. The schools may have been too attentive to verbal skills and too little attentive to the need for mechanical talents. Yet the requirement of functional literacy, at least, is built into most job expectations, and indeed into most social roles. (Marshall McLuhan thinks that writing has been made obsolete by the visual image of television, but it is worth noting that McLuhan is an inveterate producer of poorly reasoned books.)

The problem of the "unemployables" will require some social ingenuity both in altering the criteria of employability and in altering the degree and types of learning opportunities for the educational misfits. Those with advanced educational attainments are by no means totally immune to the threat of technical obsolescence in view of rapid changes in relevant knowledge and techniques. Partial immunity to the hazards of cumulative incompetence may be gained by rigging the system: tenure for teachers and professors, licensing for professional practitioners without later re-examination to determine current competence, and all sorts of seniority provisions that equate length of service with quality of performance. If all else fails, one may become an administrator of technicians, which is no small task at that. The problem of skills falling into obsolescence remains real, however, and again has implications for educational policies and procedures. First, formal education is most effective when it provides an extensive basis for continuous learning, least effective when it attempts to cram the human memory with current or already obsolete "information." Oliver Wendell Holmes noted that information, like fish, spoils quickly. Second, learning by doing may be an apt aphorism for educational procedures, but it does not follow that the mere practice of a trade or profession is an adequate basis for keeping current with what others are learning and doing. Thus continuing education of a somewhat formal character, including actual retraining for a new and more marketable specialty, is likely to become increasingly common in a society that devotes large resources to the production of new knowledge.

A word needs to be said about military technology. The demand for military weapons systems and other supplies is a principal market for nominally private manufacturers. The "military-industrial complex," as the late President Eisenhower called it, makes conspicuous the silly pompousness of the posturing representatives of industry demanding that government be kept out of business. However we have got into the current state of affairs in which we now find ourselves, the profits and jobs generated by military demand raise the awkward question about whether the American economy could in fact survive a secure peace. The military establishment itself is an employer of some people perhaps otherwise unemployable, and their employment at military tasks, together with formal in-service training, may increase their skills and employability for civilian jobs.

Military technology is one of the principal ways in which "useful" knowledge is transferred from technically advanced countries to underdeveloped countries—or, to use the more optimistic designation, newly developing ones. Military service provides the people of these countries with one of the few opportunities they have for exposure to machines and tools as well as to the discipline of complex organizations. Whether military expenditures are sensible for impoverished and grossly inefficient economies is a quite different question. Before one answers "no" to this question too quickly, one should note the role of the military in maintaining a semblance of internal order in countries having little experience in self-government and scant bases for social cohesion or common values and loyalties. Given the technically based power of the military, a principal social problem is getting the armed forces out of the game of government once they have taken a hand.

The spread of military technology illustrates a more general principle: that useful knowledge need not be everlastingly rediscovered and its applications reinvented. Modern communications make possible, even probable, the transfer of bits and pieces of technology. The adoption of entire technical systems has also been tried with varying success, for it presupposes prior or simultaneous changes in social organization of a truly revolutionary scale. (Thus the partially autonomous character of military establishments may enhance the possibility of borrowing complex systems, but by the same token it may limit wider applications to the civilian social order.) True, some systemic qualities—that is, the facts of necessary interdependence—must be taken into account if the transfer of technology is to have any close approximation to its original use. One encounters reports of alarm clocks being incorporated in Haitian *vodun* ceremonies, or, in the 1920's, of farm tractors rusting in Soviet "tractor stations" for want of fuel, parts, repair mechanics, or even drivers. A modern manufacturing plant is not only a technical system of impressive complexity but also implies a social system of at

least equal complexity: an organizational plan which specifies
tasks, relations, and lines of communication and authority; a dis-
ciplined and competent labor force which actually appears regu-
larly and on time; and an orderly array of rewards and penalties
to assure performance.

Yet the systemic character of technology is not total; whatever the salesman says, some units may be sensibly bought sep-
arately, without buying the whole package. This is true in part
because for many problems there are alternative solutions and
for many products alternative fabricating procedures. Above all,
it is not necessary to replicate the sequence of technical innovation as it originally occurred. Part of the initial order of invention
may have been accidental, and part genuinely dependent on prior
developments. But once the technique or process exists, it may
be available for adoption or adaptation without preliminaries.

To a remarkable extent there is now a world pool of technology ready for selection and use, despite valiant attempts at
secrecy or exclusive proprietorship. And at any given time, old
and new principles and applications can, and do, coexist. The
situation of the promoter of technical modernization in newly
developing countries can be likened to that of the shopper in a
supermarket. The shopper need have little or no concern about
when the product was first introduced into the array of goods
available, unless real improvements (and not just new packages)
have been made, and little or no concern about the sequence in
which the distributor acquired his inventory or put it on the
shelves, unless older items have since deteriorated. Any sensible
shopper distrusts the easy assumption that what is newer is automatically better.

The existence of technical diversity and therefore the possibility of choice combine to make decisions about technology a
practical problem for developing areas. To begin with, the general course of changes in the ways physical goods are produced
has been labor-saving. Increasingly, the bulk of manufacturing in
the economically developed countries has become "capital intensive" rather than "labor intensive." Yet capital is precisely
what poor countries are poor in, while, questions of quality

aside, they have a superabundance of labor. From an economic point of view, therefore, it would appear sensible for a developing country to adopt an "outmoded" technological package rather than the most recent one. This argument is given added support by the fact that the most sophisticated types of mechanization in the main require highly skilled workers, such as engineers and technicians in diagnosis and therapy. Since advanced training is also a form of capital accumulation, the generalization about the poverty in capital in underdeveloped countries holds in this case also. In sum, the seeming abundance of labor cannot leave questions of quality "aside" after all.

Political reality, however, intervenes. For political leaders in countries trying to catch up with the relatively rich countries, and having available the most advanced technology the world offers, it is difficult to adopt a more primitive (even if more fitting) system. Just as a military system may be a costly piece of national symbolism, so may the hydroelectric-powered aluminum fabricating plant in a world market glutted with aluminum, or the semi-automated textile plant that produces superior fabrics which the unemployed population cannot afford to buy.

The transfer of technology from one context to another has social preconditions as well as consequences: for example, its successful adoption may require a stable political order, a market system, domestically produced or imported skills, and geographical and social mobility on the part of workers independent of kinship ties. These conditions are not impossible to meet, as the air-polluting smoke from countless new factories testifies in towns previously tainted only by time-honored poverty.

Consumption and Styles of Life

Many urban consumers of milk have never seen a cow, and, except for a few elderly, displaced farmers, none would know how to milk one. (But then, probably, neither would most dairymen, since they long ago shifted to mechanical milkers.) Besides, the milk that reaches the consumer in a hard-to-open, disposable plastic carton has slight resemblance to the "natural" product issuing from the cow. It has been homogenized, irradiated, defatted or brought to a uniform butterfat content, and otherwise tampered with and thereby presumably improved. Similarly, "fresh" orange juice served in a restaurant is a substance that started as orange juice, has been dehydrated, frozen, sealed in a metal can, opened, rehydrated, aerated by mechanical stirring, and served as though just out of an orange skin.

Now these examples of the "artificiality" of consumer products are not about to set me off on a windy discourse about the virtues of the raw and simple. Wheaties may be more nutritious than plain boiled wheat, and I take it that most consumers think them less of a nuisance to prepare and certainly tastier. Perhaps only the ultimate producer and the ultimate consumer share a small unease about the fact that very little of the retail cost of most products is represented by the price paid to the farmer or other original supplier. Yet the simpler life was not necessarily a better one; on the average, it certainly was not a healthier or longer one.

The flow of goods from American factories and processing plants is truly tremendous. Commentators tend to use such terms as "flood" or "torrent" (to keep everything fluid, critics say that the consumer is being "swamped"). Aside from sheer volume, which is impressive, three further characteristics of consumer goods are noteworthy:

(1) The goods are almost entirely artificial in some degree. I do not mean anything nasty by the term, but even "garden-fresh" vegetables are commonly washed, selected, graded, packed, shipped, stored, and often repacked into smaller packages, counted or weighed, and, not least, priced. By the time they arrive at the market they are many technical stages and often a considerable distance from the good earth.

(2) Particular products are standardized in high degree, which permits mechanized mass production for a mass market.

(3) Yet the counterpart of standardization is a tremendous multiplication of products. True, some differences in brand names, packages, or claimed beneficial ingredients are distinctions without a difference. Not all, however. The variations in quantity,

price, and quality even among seemingly similar products are so great that public-spirited but sporadic attempts are made to protect the consumer both from outright deceit by the seller and from an almost impossibly complex set of decisions. Reasonably intelligent and reasonably well-informed consumers are expected to make reasonable choices. Yet reasonableness generally fails for want of valid information. Willful concealment and deceits are not the only problems. Some information, such as the chemical composition of allegedly therapeutic drugs, may be inherently unintelligible to anyone but a pharmaceutical chemist—and he may be a bit unreliable, too.

Standardization has been the target of much social criticism of a product-oriented mass of consumers. The critics exaggerate. Look-alike houses containing children who look alike regardless of age and sex can be found in considerable number, and one automobile does look very much like almost all other automobiles, manufacturers' claims to the contrary notwithstanding. But if one went through the laborious exercise of comparing all consumption goods (not to mention services) across classes of goods—say, food and furniture to various adult toys—or across classes of consumers—say, rich and poor, or urban and suburban, or executives and professionals—one would discover an astonishing diversity. True, standardization may lead to some real or spurious uniformity across such ostensibly important differences as income—witness, for example, the substantial market for plastic television antennas. Yet it is equally true that standardization speeds up the reconfirmation of a universal law of social change, the degradation of status symbols.

Those who think they see standardization everywhere and dislike it have not only observed poorly but also have missed a fundamental social characteristic of American and a number of other societies. In an affluent society, a considerable amount of consumer expenditure is discretionary—that is, subject to choice in time and type. With any substantial degree of competition among fabricators and purveyors of goods or suppliers of services, the variety of options available to the consumer may not permit him to fulfill his fondest, if rather odd, dreams, but they go far beyond any sensible list of what he really needs. In terms of consumer habits, of course, luxuries have a way of becoming comforts and then necessities: "I don't know how we got along without the third car." We have come a long way from the time-honored triumvirate of man's physical needs—food, clothing, and shelter.

Discretionary income and expenditure have further implications. Take, for example, "materialism." If, instead of a third car, the affluent consumer opts for travel, psychoanalysis, a small trust fund for a grandchild, or a college scholarship for some member of the "deserving poor," is this materialism? The equation between having or spending money and being materialistic is simply not valid. Money is useful for whatever it will buy, including eternal salvation if that market is still open. Anyone not interested in money in modern society is either being supported by those who are interested, or is certifiably insane (or possibly both).

A second issue revolving around discretionary income and expenditure especially concerns the nonmaterial results of high productivity. "Time is money," according to a time-honored maxim much favored by advocates of the active and productive life. The equation is true, of course, to the extent that time can be sold for money. I have used the phrase "discretionary income and expenditure" precisely because at least part of the working population has some choice in the allocation of time. The time freed by higher productivity may be used to produce even more, or spent in enjoying what is already available.

To a remarkable degree, when presented with an option, American workers have preferred work (and income) to leisure. Of course, leisure has a negative value for most of the unemployed, but even wage and salary earners and professionals charging fees for their services may work much more than would be necessary for mere necessities and some of life's discretionary amenities. The male wage earner "moonlights" by holding a second job, and his wife may also work (and this again may be viewed as moonlighting, since housework *is* work—it's just unpaid). The professional, aside from being normally eager for income, commonly finds his work interesting. Even the salaried professional tends to work longer hours than his employer requires.

Crowded highways on weekends, national parks, sports stadiums, beaches, trout streams, and picture galleries testify to the existence of ample residual leisure, however. Technology as commonly understood accounts both for the leisure and for many of the ways of enjoying it (or at least *using* it). Contemplate the home workshop with its power tools, the collection of cameras and perhaps darkroom equipment of the photography buff, the elaborate gear of the golfer or fisherman, the assembly of components for a stereophonic high-fidelity system. The mind boggles and the producers profit. Do the consumers?

Well, why not? The critics of "mass culture" are simply not talking about anything real if their assumption is that standardization is not merely pervasive but always bad. The problem goes deeper. The attitude of the critics seems to me to be snobbish if not downright snotty. Is Beethoven less stirring, Tchaikovsky less melodic, or Bach less mathematically elegant if selected and heard by a grocery clerk who has a collection of classics? Does the enigmatic smile of the Mona Lisa lose its interest when reproduced in a faithful print that hangs on the schoolteacher's apartment wall? Is the dockhand's wife getting out of place if her front parlor is graced by furniture out of Grand Rapids in the French Provincial style? Since the critics of "mass culture" are commonly from either the Old Left or the New Left, one can only wonder why they dislike the extension of the benefits of civilization to the (mythical) masses.

Who's in Charge Here?

The doctrine that holds technology to be the primary factor of all social causation does not need yet another post-mortem lethal blow. But some issues remain to be explored and clarified. The question is not whether technology causes social change: it does;

or whether various social changes cause technology: they do. The only interesting question is: Which changes under what circumstances?

Nearly all of current technical innovation is deliberate. This means, concretely, that technological change is organized and institutionalized. The "pure" knowledge that is expanded and the useful knowledge that is applied are almost entirely a function of the amount and type of support given to some fields and projects, and thus not to others. There is nothing in abstract or applicable knowledge itself that would account for its expansion or use without human intervention.

In a "technocracy"—that idle dream of technologists wishing to be masters of their own destiny as well as that of others— the decisions about the flow of funds would be made by their initial users. We should have to be naive to the point of idiocy to suppose that the technocrats would abandon their loyalties to their specialties, their conscientious belief in the superiority of their solutions among the competitive alternatives. So far, much in-fighting otherwise to be expected has been controlled by enlarging the size of the pie to be cut, but that does not quiet those who get the smaller pieces. There are as many actual or potential technologies as there are problems to be solved or alleviated, and without unlimited resources the technologists are bound to compete for their particular pet projects.

The plurality of technologies warrants underscoring, for useful knowledge is not confined to mechanical, chemical, or biological principles. There is also social technology. It is no more ascendant or final than any other technology. All technology is instrumental, and the uses to which useful knowledge is to be directed are determined by social values and preferences, by organizations and other coalitions of interests.

The current quest for the control of technology is aimed primarily at the unforeseen, negative consequences of intervention in the "natural" order of things. The time and energy increasingly devoted to "technological assessment" represent an attempt to predict the secondary and tertiary effects of a technical innovation and, taking the realistically pessimistic view that these consequences are likely to be negative—that is, very costly—the task of assessment is to weigh the costs against the prospective primary benefits. The assembly of teams of experts to engage in technological assessment is thus itself an interesting bit of social technology.

Other instruments of social control are available, most notably the legislative, administrative, and even judicial processes afforded by law. These still tend to be remedial rather than preventive. The ultimate control of technology, including harmful applications of what we could do, rests on complex political processes for registering and compromising human purposes. This does not put the whole matter beyond our ken and control. I believe that it does clarify much muddy water. For example, it does not make the technologists the sole or primary perpetrators of our environmental or other ills. They only stand properly accused of being pure instrumentalists, of being men of little vision or at least not expressing it. ("Theirs not to question why, theirs but to cut and try.") Above all, the political focus converts technological change from a spurious given, a datum, into a real problem. And that is very close to where we came in.

Part 1

THE CONSEQUENCES OF TECHNOLOGICAL CHANGE

MIXED REACTIONS to a rapidly changing technology are probably older than recorded history. We do not know how the wheel, the bow and arrow, and the earthen pot were initially greeted, but it would be surprising if there were not some scoffers present. Since it is not necessarily true that any change is an improvement, and given human ingenuity in concocting silly contraptions, some inventions have deserved distrust on their own merits.

Recent and current criticism of technology rarely objects to particular innovations (though witness the possibly reasonable hostility to the eardrum-splitting, window-smashing, supersonic transport). More frequently, the unequal pace of technology in various fields, or the presumed lag in social adaptation to new products and techniques, or such social costs as pollution invite the attention of critical commentators.

The first two essays in this section comprise a complementary pair of generally optimistic but cautionary views of the pace of technical change. Both William Ogburn and Bertrand Russell profess to be interested in the social implications of science, but turn out to be concerned with applied science or technology. This is just as well, for science as such has few social implications, and technology has many. Both the sociologist Ogburn and the philosopher-mathematician (and inveterate pleader for both noble and silly public causes) Russell plead for social reason in the use of technology, and imply that technology leads the way, with social arrangements lagging behind. Yet more than two decades ago both men expressed some hope for man's capacity to cope with his own handiwork; neither suggested a simple moratorium on technical change while other forces catch up.

Kenneth Keniston's short essay also indulges in the "cultural lag" type of reasoning: technology has outpaced social adaptation. In Keniston's view, the power of technology is poorly contained by a rather backward set of ethical principles. One may raise the question, however, as to when ethics might intervene: why not as a preventive restraint rather than a merely "coping" response? In no logical, grammatical, or ethical system are "can" and "should" equivalent. Anything we can do we can also decide not to do.

The essays by A. H. Raskin, Eric Hoffer, and David Dempsey may also be viewed as complementary. Raskin is not unmindful of the problems caused by the displacement of men by machines, but he perceptively discusses the machine "mastery" granted to those left behind at the factory. The notion of work as creative rather than solely burdensome was formerly associated only with artistic, entrepreneurial, and professional types. Its extension to at least part of the workers more closely involved in the production of ordinary consumer goods means that what used to be a common distinction between white-collar and blue-collar jobs is further impaired. Hoffer attends both to the abundance of material goods made possible by late stages in mechanization and to the question of what to do with the time made available by productive efficiency. Leisure (properly understood as "discretionary time") may mean, he says, mere inaction ("time on one's hands") or frivolous and even dangerous action. Dempsey, too, is worried about leisure. He is discerning with respect to the round of activities that absorb seemingly residual time, though I think he exaggerates their obligatory character. Filling up time may be no more than an excuse for the failure to spend leisure creatively. Perhaps most of us have not yet escaped a kind of Calvinistic work ethic, whereby the only justifiable life is the active one: work hard and play hard (and always be useful).

Unwanted leisure has long been the lot of farmers in temperate zones, owing to the seasonality of productive labor. Vernon Vine reminds us that the Industrial Revolution has also hit agriculture, with resulting multiplication of productivity per unit of labor, but also greater output per unit of land. One small fallout from the changing character of agriculture is the threatened position of a once-prosperous scholarly pursuit: rural sociology. Rural America is not radically different from urban America, only a bit less congested. Changing technology sends former farmers and farm-hands to the cities, and rural sociologists abroad to India or Latin America.

No set of essays on the consequences of technological change would be complete without some reference to the apostle of televised communication, Marshall McLuhan. Since, according to McLuhan, the "medium is the message," and since most of his intended messages have been conveyed in printed writing (with its lineal restraints), it is perhaps understandable that McLuhan impresses the archaic logical mind as a better phrase-maker than he is a coherent thinker. I have therefore taken McLuhan secondhand by way of a friendly essay by Richard Kostelanetz. Kostelanetz notes that McLuhan denies being a technological determinist, preferring the characterization of an "organic auton-omist," whatever that may mean. McLuhan does argue, however, that the mode of communication affects our attitudes and even our perceptions. He even argues, according to the news report by John Leo, that "aimless violence" is due to television, not because of the imitation of gunfights between bad guys and good guys, but because adolescents hate machines and cities and are involved in an identity quest. If television won't keep the viewer out of mischief but rather promotes destructive upheavals, where can we turn next? Lobotomies, perhaps, or massive doses of tranquilizers forcibly administered? Or does reason still have a chance, McLuhan to the contrary notwithstanding?

Can Science Bring
Us Happiness?

by William F. Ogburn

I ONCE amused myself by asking persons on the street, in stores
and on street cars what they most desired out of life. This was
before the polls of Mr. Gallup or of the inquiring reporter, or
the stratified sample. I expected those questioned to place foremost good health, which Herbert Spencer ranked first. But they
did not. Neither did they say salvation or life after death. Quite
differently, their first wish was more money to spend. Secondly,
they wanted more happiness out of life.

These may not be the answers I thought they ought to give,
but they appeared to be sincere and honest answers. Who does
not want happiness, or more money to spend? As these seem to
be almost universal desires they merit consideration. So, since
taking this poll I have from time to time speculated on what
social science might do to bring about these desiderata.

Our first question is: What is the promise of social science in
regard to the standard of living? After examining that, we can
consider what social science promises in regard to happiness.

The outlook for a higher plane of living for all of us here in

From the *New York Times Magazine*, December 4, 1949, copyright ©
1949 by The New York Times Company.

the United States is very good indeed. The wealth we probably will have, however, will not be wholly an achievement of social science; it will be the result also of natural science. For we must have tools as well as markets to produce the wealth.

In truth, the standard of living of a people is a function of four factors—population, natural resources, inventions, and social organization, all working together. Let us see what part each plays.

Population: For a nation as a whole there are points beyond which a larger population means a smaller per capita income if there is no new technology. In China the average farm is about four acres, and the living standard is low. If the population of China grows larger without improvements in science and invention, the average income per farmer will fall even lower.

In densely peopled Italy, on the other hand, population increase between the two world wars was accompanied by an increasing rather than a decreasing per capita income. How could this be? The answer is that the advance in technology and the improvement in economic organization overcame the depressing effect of population increase.

A high standard of living is better assured with a not-too-large population in relation to land area. For the United States, in the next half century, the probabilities are that we shall have a relatively small population for our territory, perhaps 175,000,000 or 200,000,000. Such totals are indicated by projecting past birth and death rates to successive years of population estimates. These rates indicate a continued decline in birth rates sufficient to offset the decline in death rates so as to reach the indicated figures.

If the estimates are sound, the United States, from the population standpoint alone, seems assured of a rising living standard.

Natural resources: With regard to agriculture, science shows us how the fertility of the soil may be maintained even as to trace minerals. In industry natural resources are also important. There the variety used is so great that highly industrialized countries must draw upon the mineral resources of the world. As technology develops, the demand increases for many metals such as iron, bauxite, nickel, tin, cobalt, titanium, uranium and thorium. Thus we may expect rising prices.

In addition to such natural resources, mechanical power is needed for a high living standard. To coal, oil and falling water, we are to add the power that comes from the atom. The expectation is that the atom will produce power in abundance.

Thus, considering this country's natural resources and power resources, it is probable that the United States will be able to support a very much higher living standard for its people.

Inventions: The most important influences in raising the standard of living of modern times, and indeed of all times, are scientific progress and the invention of mechanical devices. A hundred years ago the average hourly output of a laborer in the United States was valued at 27 cents. Today the laborer's average hourly production is worth $1.32, or five times as much. Yet he is not five times as strong, nor does he work five times as hard. The answer to his increased productivity lies, in part, in his use of better tools and of mechanical power. The latter accounts for 92 per cent of the energy used in our present wonderful productivity.

An impressive observation about inventions is that they keep coming. New patents are applied for at the rate of about 80,000 a year. The effect of this onward surge of inventions can be seen within the lifetime of any older person. When I was a boy telephones were not common. Automobiles were so new that President Theodore Roosevelt was praised editorially for his courage in riding in one. The airplane was considered a crank's dream, and Simon Newcomb, dean of scientists, questioned whether man would ever fly in a heavier-than-air machine.

The youth of 1949 in the next fifty years will see even more wonderful inventions than I have seen. We do not know what they will be, though we may guess at some of them. But whatever they are, our standard of living will be enriched by them, and we shall be able to produce with fewer hours of work and with less expenditure of human energy.

Social organization: This fourth factor in the standard of living is of utmost importance in a complex, industrial civilization. In the Nineteen Thirties, this country had plenty of factories and machines, lots of coal and iron in the ground, and many people who wanted to work. Yet factories and resources and people could not get work. It requires more than labor and inventions and natural resources to produce. The economic organization must function too.

There is now much concern about the future of our economic organization. Will our efficient private capitalism under which the standard of living has increased so marvelously be superseded by a perhaps less efficient socialism? There is no doubt that much of the world is moving toward some sort of union of government and industry, toward the ownership by government of large essential industries. There is much talk, indeed, of utter collapse through the breakdown of the capitalist system in a vast depression, with Communist-encouraged chaos in the offing.

This talk is wild. History affords few examples of collapse and chaos. There was a French Revolution and a Russian Revolution, but productive economic organizations survived them both. There is now a Chinese Revolution, which also will be followed by a producing social mechanism. Empires have fallen, but seldom into complete chaos unless there was war. Wherever human society exists there is social organization, and nearly always there is a pretty good adjustment to the tools of production.

Nevertheless, there can be very serious setbacks. The most destructive force, at least temporarily, is war. With the application of science to destruction on the scale now possible, our social organization might suffer severely from a war within the next half century. Even without war there are dangers. The science of economics does not yet have a precise answer as to how efficient or inefficient nationalization of industry will be, despite the fact that there are strong convictions on the subject. Nor do we definitely know in practice how severe depressions may be avoided.

The effectiveness of measures to guard against these dangers depends somewhat on knowledge of social science, but much more on the application of this knowledge. It is held, for instance, that spending properly in a depression and paying back the expenditures out of taxes in periods of prosperity will lessen the swings of the business cycle. But the trouble lies in application of this theory—namely, in getting the people to tax themselves enough in prosperity to pay back what was borrowed in hard times.

Besides the conclusions to be drawn from the four major factors, there is another approach to the question of the future living standard. This is to project past trends forward. One trend that we can project is income—not total income but per capita income, and not income in prices of the time but in terms of real purchasing power.

For a half century preceding World War II, the national per capita income in money of the same purchasing power increased at the average rate of 1.6 per cent a year, a historical fact of considerable importance. This increase means that the standard of living actually more than doubled from 1889 to 1939. If the amount of money the average person had to spend has more than doubled in the past fifty years, it is not unreasonable to think that it may do so again in the next fifty years.

We do not know, of course, that income will increase at that average rate. It may increase at a faster rate, as it has during certain periods of the past fifty years. Or at a lower rate, as it has at other periods. We observe that acceleration has been the rule rather than the exception in the past century. On this assumption we might expect a higher rate of growth than 1.6 per cent a year. On the other hand, it is safer to expect a low average rate of increase. However, if it does no more during the next fifty years than double the present living standard, it will remain a phenomenon to be marveled at.

But despite this optimistic outlook we must, realistically, recognize one prime danger—possible failure of social organization, which includes political, economic and other types of organization. Prolonged disorganization, whether resulting from a destructive war or other causes, could ruin the chances of a much higher standard of living. The trend toward bigness and complexity also presents a danger in this connection.

The indications are that the future will see a much more highly organized society than we have ever had before. Intricate as is the present organization of industrialism, there is much more of it to come; some kind of union of government and industry is to be expected, and both are expanding more and more elaborately. The tremendous organizations in prospect for the future lay a heavy burden on democracy, if we define democracy as government by the people. For it will not be as easy for the people to govern such huge organizations as it was to govern a New England village through the town meeting. Furthermore, the problem of liberty will be made more acute.

The chief danger to the living standard, however, lies in the fact that in large and intricate mechanisms, such as government and industry are likely to become, considerable disorganization may occur from what appear to be relatively slight causes. It is like a factory that can be put out of operation by a lack of ball bearings or the breakdown of one key machine.

On the other hand, social science, now very young, is likely to make rapid advances during the next century. As it does so, efficiency will be increased. Figures presented by Frederick Mills of the National Bureau of Economic Research show that the income from labor output virtually doubled from 1899 to 1947. The additional income is due to increased efficiency through organization and technology.

In the future the growing use of statistics by social science and of quantitative measurement will surely tell us more about the influences that produce irregularities in the flow of production and purchasing power and so cause depressions.

Thus, while there is danger to an increasing standard of living from possible disruptions in our intricate social organization, there is a good prospect that other factors will overcome the danger. The factor most favorable to a higher living standard is that of new inventions.

For the second great hope of mankind—more happiness during our brief sojourn on this planet—the promise of social science is not so definite. My poll showed that most people wanted happiness, but I am not sure most of them could have defined happiness. It is not the same thing as material welfare, for many rich people are unhappy. Nevertheless, there is some correlation between the two. It is difficult to see how a destitute population, lacking enough good food, could be happy much of the time.

There are two items from nutrition that are said to be related to the state of pleasurable well-being which is called happiness. One is calcium, the regulation of which is the function of the parathyroids. It is believed that with bad functioning of the parathyroids and a deficiency of calcium, it is more difficult for a person to be happy. There is no theory, however, which claims that the retention of enough calcium will bring happiness.

A similarly negative bit of evidence comes from thiamine. Several very convincing experiments have been performed which have shown that a marked absence of Vitamin B-1 from the diet for several weeks produces a state of apathy, listlessness and lack of buoyancy. The low morale accompanying a deficiency of thiamine has given this vitamin the popular name of "morale vitamin." But here again the presence of this factor is not a promise of happiness, although its marked absence is almost a guarantee of unhappiness.

Looking elsewhere for the meaning of happiness, consider what classes of people are distinctly unhappy. Among them are the neurotic and psychotic. Neurotics suffer from emotional conflicts which come chiefly from distortions of sex and maladjustments of the ego. Unhappiness results also from the loss of a loved one, and estranged married couples usually go through a period of unhappiness.

Thus, while nutritional and other factors may have a great deal to do with unhappiness, it would seem that the social institution most closely associated with happiness or unhappiness is the family. We say a happy home, or a happy marriage. We refer to a happy or an unhappy childhood. The most common elements in the various institutions and situations that bring happiness comprise the sentiments related to affection, and these center in marriage, family, the home.

Next in importance in this respect I would rate either economic institutions that provide one's work, or the church. There are some who find happiness, they say, in their work. "Philosophy is my mistress," says the philosopher. Yet for most people this seems to be an ego satisfaction rather than a real affection. To others, religion is a source of peace and ecstasy that might be called happiness.

Social science, then, in putting itself to the service of advising on how more happiness may be had, focuses on the family. While there may be other contributing institutions, the home is the best point of attack.

But does not the rising divorce rate rather dim any prospective promises of social science for a happier people? Not necessarily. There were unhappy marriages when divorces were negligible in number, just as there was sickness before there were hospitals. Divorced persons remarry, and often second marriages are happier than first marriages. Furthermore, many surveys of sociologists show in the samples of families from the United States that some 80 or 85 per cent of all marriages are happy.

It is only recently that scientific study of the family has been undertaken, and the progress is noteworthy. Research in the social psychology of personality formation and adjustment is also young, yet already there are many hundreds of courses on the family and marriage in the colleges and high schools.

As personality difficulties in the family are studied, the causes of mismatings are more and more traced to early childhood patterns. Hence the rearing of babies and children as the basic foundation on which to lay a program for a happier family life and for more happiness in general becomes of prime importance.

Social science does not yet have a great deal of achievement to point to in this field. But there are many good studies that have developed probable hypotheses. Surely the promise is good. Fifty or a hundred years hence we should have much more exact information than we now have. A greater concern for the problems of personality in social engineering will be needed as well. For even after social science develops the knowledge, it must be applied by millions at home and by millions of schoolteachers.

Biblical tradition has handed down to us the statement: ". . . the poor ye have always with you"; and our mundane habitation has been characterized as a "vale of tears." But now we are about to see poverty abolished in our own country, if we do not act foolishly. And there is reason to think that we may yet be a happier people.

The Science to Save
Us from Science

by Bertrand Russell

SINCE THE BEGINNING of the seventeenth century scientific discovery and invention have advanced at a continually inc‗.asing rate. This fact has made the last three hundred and fifty years profoundly different from all previous ages. The gulf separating man from his past has widened from generation to generation, and finally from decade to decade. A reflective person, meditating on the extinction of trilobites, dinosaurs and mammoths, is driven to ask himself some very disquieting questions. Can our species endure so rapid a change? Can the habits which insured survival in a comparatively stable past still suffice amid the kaleidoscopic scenery of our time? And, if not, will it be possible to change ancient patterns of behavior as quickly as the inventors change our material environment? No one knows the answer, but it is possible to survey probabilities, and to form some hypotheses as to the alternative directions that human development may take.

The first question is: Will scientific advance continue to grow

From the *New York Times Magazine,* March 19, 1950, copyright © 1951 by The New York Times Company, with permission of the Estate of Bertrand Russell.

more and more rapid, or will it reach a maximum speed and then begin to slow down?

The discovery of scientific method required genius, but its utilization requires only talent. An intelligent young scientist, if he gets a job giving access to a good laboratory, can be pretty certain of finding out something of interest, and may stumble upon some new fact of immense importance. Science, which was still a rebellious force in the early seventeenth century, is now integrated with the life of the community by the support of governments and universities. And as its importance becomes more evident, the number of people employed in scientific research continually increases. It would seem to follow that, so long as social and economic conditions do not become adverse, we may expect the rate of scientific advance to be maintained, and even increased, until some new limiting factor intervenes.

It might be suggested that, in time, the amount of knowledge needed before a new discovery could be made would become so great as to absorb all the best years of a scientist's life, so that by the time he reached the frontier of knowledge he would be senile. I suppose this may happen some day, but that day is certainly very distant. In the first place, methods of teaching improve. Plato thought that students in his academy would have to spend ten years learning what was then known of mathematics; nowadays any mathematically minded schoolboy learns much more mathematics in a year.

In the second place, with increasing specialization, it is possible to reach the frontier of knowledge along a narrow path, involving much less labor than a broad highway. In the third place, the frontier is not a circle but an irregular contour, in some places not far from the center. Mendel's epoch-making discovery required little previous knowledge; what it needed was a life of elegant leisure spent in a garden. Radio-activity was discovered by the fact that some specimens of pitchblende were unexpectedly found to have photographed themselves in the dark. I do not think, therefore, that purely intellectual reasons will slow up scientific advances for a very long time to come.

There is another reason for expecting scientific advance to continue, and that is that it increasingly attracts the best brains. Leonardo da Vinci was equally pre-eminent in art and science, but it was from art that he derived his greatest fame. A man of similar endowments living at the present day would almost certainly hold some post which would require his giving all his time to science; if his politics were orthodox, he would probably be engaged in devising the hydrogen bomb, which our age would consider more useful than his pictures. The artist, alas, has not the status that he once had. Renaissance princes might compete for Michelangelo; modern states compete for nuclear physicists.

There are considerations of quite a different sort which might lead to an expectation of scientific retrogression. It may be held that science itself generates explosive forces which will, sooner or later, make it impossible to preserve the kind of society in which science can flourish. This is a large and different question, to which no confident answer can be given. It is a very important question, which deserves to be examined. Let us therefore see what is to be said about it.

Industrialism, which is in the main a product of science, has provided a certain way of life and a certain outlook on the world. In America and Britain, the oldest industrial countries, this outlook and this way of life have come gradually, and the population has been able to adjust itself to them without any violent breach of continuity. These countries, accordingly, did not develop dangerous psychological stresses. Those who preferred the old ways could remain on the land, while the more adventurous could migrate to the new centers of industry. There they found pioneers who were compatriots, who shared in the main the general outlook of their neighbors. The only protests came from men like Carlyle and Ruskin, whom everybody at once praised and disregarded.

It was a very different matter when industrialism and science, as well-developed systems, burst violently upon countries hitherto ignorant of both, especially since they came as something foreign, demanding imitation of enemies and disruption of ancient national habits. In varying degrees this shock has been endured by Germany, Russia, Japan, India and the natives of Africa. Everywhere it has caused and is causing upheavals of one sort or another, of which as yet no one can foresee the end.

The earliest important result of the impact of industrialism on Germans was the Communist Manifesto. We think of this now as the Bible of one of the two powerful groups into which the world is divided, but it is worth while to think back to its origin in 1848. It then shows itself as an expression of admiring horror by two young university students from a pleasant and peaceful cathedral city, brought roughly and without intellectual preparation into the hurly-burly of Manchester competition.

Germany, before Bismarck had "educated" it, was a deeply religious country, with a quiet, exceptional sense of public duty. Competition, which the British regarded as essential to efficiency, and which Darwin elevated to an almost cosmic dignity, shocked the Germans, to whom service to the state seemed the obviously right moral ideal. It was therefore natural that they should fit industrialism into a framework of nationalism or socialism. The Nazis combined both. The somewhat insane and frantic character of German industrialism and the policies it inspired is due to its foreign origin and its sudden advent.

Marx's doctrine was suited to countries where industrialism was new. The German Social Democrats abandoned his dogmas when their country became industrially adult. But by that time Russia was where Germany had been in 1848, and it was natural that Marxism should find a new home. Stalin, with great skill, has combined the new revolutionary creed with the traditional belief in "Holy Russia" and the "Little Father." This is as yet the most notable example of the arrival of science in an environment that is not ripe for it. China bids fair to follow suit.

Japan, like Germany, combined modern technique with worship of the state. Educated Japanese abandoned as much of their ancient way of life as was necessary in order to secure industrial and military efficiency. Sudden change produced collective hysteria, leading to insane visions of world power unrestrained by traditional pieties.

These various forms of madness—communism, nazism, Japanese imperialism—are the natural result of the impact of science on nations with a strong pre-scientific culture. The effects in Asia are still at an early stage. The effects upon the native races of Africa have hardly begun. It is therefore unlikely that the world will recover sanity in the near future.

The future of science—nay more, the future of mankind—depends upon whether it will be possible to restrain these various collective hysterias until the populations concerned have had time to adjust themselves to the new scientific environment. If such adjustment proves impossible, civilized society will disappear, and science will be only a dim memory. In the Dark Ages science was not distinguished from sorcery, and it is not impossible that a new Dark Ages may revive this point of view.

The danger is not remote; it threatens within the next few years. But I am not now concerned with such immediate issues. I am concerned with the wider question: Can a society based, as ours is, on science and scientific technique, have the sort of stability that many societies had in the past, or is it bound to develop explosive forces that will destroy it? This question takes us beyond the sphere of science into that of ethics and moral codes and the imaginative understanding of mass psychology. This last is a matter which political theorists have quite unduly neglected.

Let us begin with moral codes. I will illustrate the problem by a somewhat trivial illustration. There are those who think it wicked to smoke tobacco, but they are mostly people untouched by science. Those whose outlook has been strongly influenced by science usually take the view that smoking is neither a vice nor a virtue. But when I visited a Nobel works, where rivers of nitro-glycerine flowed like water, I had to leave all matches at the entrance, and it was obvious that to smoke inside the works would be an act of appalling wickedness.

This instance illustrates two points: first, that a scientific outlook tends to make some parts of traditional moral codes appear superstitious and irrational; second, that by creating a new environment science creates new duties, which may happen to coincide with those that have been discarded. A world containing hydrogen bombs is like one containing rivers of nitro-glycerine; actions elsewhere harmless may become dangerous in the highest degree. We need therefore, in a scientific world, a somewhat different moral code from the one inherited from the past. But to give to a new moral code sufficient compulsive force to restrain actions formerly considered harmless is not easy, and cannot possibly be achieved in a day.

As regards ethics, what is important is to realize the new dangers and to consider what ethical outlook will do most to diminish them. The most important new facts are that the world is more unified than it used to be, and that communities at war with each other have more power of inflicting mutual disaster than at any former time. The question of power has a new importance. Science has enormously increased human power, but has not increased it without limit. The increase of power brings an increase of responsibility; it brings also a danger of arrogant self-assertion, which can only be averted by continuing to remember that man is not omnipotent.

The most influential sciences, hitherto, have been physics and chemistry; biology is just beginning to rival them. But before very long psychology, and especially mass psychology, will be recognized as the most important of all sciences from the standpoint of human welfare. It is obvious that populations have dominant moods, which change from time to time according to their circumstances. Each mood has a corresponding ethic. Nelson inculcated these ethical principles on midshipmen: to tell the truth, to shoot straight, and to hate a Frenchman as you would the devil. This last was chiefly because the English were angry with France for intervening on the side of America. Shakespeare's Henry V says:

> *If it be a sin to covet honor,*
> *I am the most offending soul alive.*

This is the ethical sentiment that goes with aggressive imperialism: "honor" is proportional to the number of harmless people you slaughter. A great many sins may be excused under the name of "patriotism." On the other hand complete powerlessness suggests humility and submission as the greatest virtues; hence the vogue of stoicism in the Roman Empire and of Methodism among the English poor in the early nineteenth century. When, however, there is a chance of successful revolt, fierce vindictive justice suddenly becomes the dominant ethical principle.

In the past, the only recognized way of inculcating moral precepts has been by preaching. But this method has very definite limitations: it is notorious that, on the average, sons of clergy are not morally superior to other people. When science has mastered this field, quite different methods will be adopted. It will be known what circumstances generate what moods, and what moods incline men to what ethical systems. Governments will then decide what sort of morality their subjects are to have and their subjects will adopt what the Government favors, but will do so under the impression that they are exercising free will. This may sound unduly cynical, but that is only because we are not yet accustomed to applying science to the human mind. Science has powers for evil, not only physically, but mentally: the hydrogen bomb can kill the body, and government propaganda (as in Russia) can kill the mind.

In view of the terrifying power that science is conferring on governments, it is necessary that those who control governments should have enlightened and intelligent ideals, since otherwise they can lead mankind to disaster.

I call an ideal "intelligent" when it is possible to approximate to it by pursuing it. This is by no means sufficient as an ethical criterion, but it is a test by which many aims can be condemned. It cannot be supposed that Hitler desired the fate which he brought upon his country and himself, and yet it was pretty certain that this would be the result of his arrogance. Therefore the ideal of "Deutschland ueber Alles" can be condemned as unintelligent. (I do not mean to suggest that this is its only defect.) Spain, France, Germany and Russia have successively sought world dominion: three of them have endured defeat in consequence, but their fate has not inspired wisdom.

Whether science—and indeed civilization in general—can long survive depends upon psychology, that is to say, it depends upon what human beings desire. The human beings concerned are rulers in totalitarian countries, and the mass of men and women in democracies. Political passions determine political conduct much more directly than is often supposed. If men desire victory more than cooperation, they will think victory possible.

But if hatred so dominates them that they are more anxious to see their enemies killed than to keep their own children alive, they will discover all kinds of "noble" reasons in favor of war. If they resent inferiority or wish to preserve superiority, they will have the sentiments that promote the class war. If they are bored beyond a point, they will welcome excitement even of a painful kind.

Such sentiments, when widespread, determine the policies and decisions of nations. Science can, if rulers so desire, create sentiments which will avert disaster and facilitate cooperation. At present there are powerful rulers who have no such wish. But the possibility exists, and science can be just as potent for good as for evil. It is not science, however, which will determine how science is used.

Science, by itself, cannot supply us with an ethic. It can show us how to achieve a given end, and it may show us that some ends cannot be achieved. But among ends that can be achieved our choice must be decided by other than purely scientific considerations. If a man were to say, "I hate the human race, and I think it would be a good thing if it were exterminated," we could say, "Well, my dear sir, let us begin the process with you." But this is hardly argument, and no amount of science could prove such a man mistaken.

But all who are not lunatics are agreed about certain things: That it is better to be alive than dead, better to be adequately fed than starved, better to be free than a slave. Many people desire those things only for themselves and their friends; they are quite content that their enemies should suffer. These people can be refuted by science: Mankind has become so much one family that we cannot insure our own prosperity except by insuring that of everyone else. If you wish to be happy yourself, you must resign yourself to seeing others also happy.

Whether science can continue, and whether, while it continues, it can do more good than harm, depends upon the capacity of mankind to learn this simple lesson. Perhaps it is necessary that all should learn it, but it must be learned by all who have great power, and among those some still have a long way to go.

Does Human Nature Change in a Technological Revolution?

by Kenneth Keniston

SINCE THE BEGINNING of terrestrial history, man has been subject to nature. Last year ended with a feat that symbolizes the central revolution of our time: man's growing control over nature. Men have for the first time sped beyond the gravity of this earth, looked down upon our distant, spinning globe and asked in sudden mockery if this planet harbored life.

This escape from the pull of the earth symbolizes a revolution in human life and consciousness as profound as any before it. The Copernican revolution removed man from the center of the universe, and the Darwinian revolution deprived him of his unique position as not-an-animal. Today the technological revolution is depriving man of both the security and the constraint that came from subjugation to nature in its given visible forms.

Man is learning to understand the inner processes of nature, to intervene in them and to use his understanding for his own purposes, both destructive and benign. Increasingly, the old reins are off, and the limits (if any) of the future remain to be defined.

But what of man himself? Do the constraints of human nature still apply? How will the change in man's relationship to nature change man? No one can answer these questions with assurance, for the future of humanity is not predestined but created by human folly and wisdom. Yet what is already happening to modern men can provide some insight into our human future.

If there is any one fact that today unites all men in the world, it is adaptation to revolutionary change in every aspect of life—in society, in values, in technology, in politics and even in the shape of the physical world. In the underdeveloped world, just as in the industrialized nations, change has encroached upon every stable pattern of life, on all tribal and traditional values, on the structure and functions of the family and on the relations between the generations.

Furthermore, every index suggests that the rate of change will increase up to the as yet untested limits of human adaptability. Thus, man's relationship to his individual and collective past will increasingly be one of dislocation, of that peculiar mixture of freedom and loss that inevitably accompanies massive and relentless change.

As the relevance of the past decreases, the present—all that can be known and experienced in the here-and-now—will assume even greater importance. Similarly, the gap between the generations will grow, and each new generation will feel itself compelled to define anew what is meaningful, true, beautiful and relevant, instead of simply accepting the solutions of the past. Already today, the young cannot simply emulate the parental generation; tomorrow, they will feel even more obliged to criticize, analyze, and to reject, even as they attempt to re-create.

A second characteristic of modern man is the prolongation of psychological development. The burgeoning technology of the highly industrialized nations has enormously increased opportunities for education, has prolonged the postponement of adult responsibilities and has made possible an extraordinary continuation of emotional, intellectual and ethical growth for millions of children and adolescents.

In earlier eras, most men and women assumed adult responsibilities in childhood or at puberty.

Today, in the advanced nations, mass education continues through the teens and for many, into the twenties. The extension of education, the postponement of adulthood, opens new possibilities to millions of young men and women for the development of a degree of emotional maturity, ethical commitment and intellectual sophistication that was once open only to a tiny minority. And in the future, as education is extended and prolonged, an ever larger part of the world population will have what Santayana praised as the advantages of a "prolonged childhood."

This will have two consequences. First, youth, disengaged from the adult world and allowed to question and challenge, can be counted on to provide an increasingly vociferous commentary on existing societies, their institutions and their values. Youthful unrest will be a continuing feature of the future.

Second, because of greater independence of thought, emotional maturity and ethical commitment, men and women will be more complex, more finely differentiated and psychologically integrated, more subtly attuned to their environments, more developed as people. Perhaps the greatest human accomplishment of the technological revolution will be the unfolding of human potentials heretofore suppressed.

Finally, in today's developed nations we see the emergence of new life styles and outlooks that can be summarized in the concept of technological man. Perhaps here the astronauts provide a portent of the future. Studies of the men who man the space capsules speak not of their valor, their dreams, or their ethical commitments, but of their "professionalism and feeling of craftsmanship," their concern "with the application of thought to problems solvable in terms of technical knowledge and professional experience," and their "respect for technical competence."

The ascendancy of technological man is of course bitterly resisted. The technological revolution creates technological man but it also creates two powerful reactions against the technological life style. On the one hand it creates, especially in youth, new humanist countercultures devoted to all that technological man minimizes: feeling, intensity of personal relationships, fantasy, the exploration and expansion of consciousness, the radical reform of existing institutions, the furtherance of human as opposed to purely technical values.

On the other hand, technological change creates reactionary counterforces among those whose skills, life styles and values have been made obsolete. In the future, the struggle between these three orientations—technological, humanistic and reactionary—will inevitably continue. Technological man, like the technology he serves, is ethically neutral. The struggle for the social and political future will therefore be waged between those who seek to rehumanize technology and those who seek to return to a romanticized, pre-technological path.

Much of what will happen to men and women in the future is good—or if not good, then at least necessary. Yet it may not be good enough. The revolution over nature has already given men the capacity to destroy tenfold all of mankind, and that capacity will be vastly multiplied in the future. And many of the likely future characteristics of men—openness, fluidity, adaptability, professional competence, technical skill and the absence of passion—are essentially soulless qualities. They can equally be applied to committing genocide, to feeding the starving, to conquering space or to waging thermonuclear war.

Such qualities are truly virtuous only if guided by an ethic that makes central the preservation and unfolding of human life and that defines "man" as any citizen of this spinning globe. So far, the technological revolution has neither activated nor extended such an ethic.

Indeed, I sometimes feel that we detect no life on any of the myriad planets of other suns in distant galaxies for just this reason. I sometimes fear that creatures on other planets, having achieved control of nature but lacking an overriding devotion to life, ended by using their control of nature to destroy their life.

In this regard, the future of man remains profoundly uncertain.

Pattern for Tomorrow's Industry?

by A. H. Raskin

DEARBORN, MICH.

IN MICHIGAN automobile factories, Illinois railroad yards, Pennsylvania oil refineries and New York brokerage offices, a new kind of industrial magic is making old operating methods look like slow motion. Its name is automation, and its ability to edit man out of the productive process is an awesome thing to watch, whether the proving ground is an insurance company's record-stuffed headquarters on Madison Avenue or the mighty River Rouge plant of the Ford Motor Company, cradle of mass production.

Here at the Rouge, in a clanking world of colored lights, block-long machines with robot controls, and noises that outroar the Times Square subway station at rush hour, the pattern for tomorrow's industry is being hammered out along with the 5,500 Ford engines that march down the assembly line each day.

The same pattern is taking shape in scores of other plants and offices all over the country. Even now, in its infancy, automation is being used to roll endless sheets of steel, make artillery

From the *New York Times Magazine,* December 18, 1955, copyright © 1955 by The New York Times Company.

shells, bake cakes, control store inventories, put through transcontinental telephone calls, compute stock margins, load freight trains, assemble television sets, design new automated equipment and speed production of the 8,000,000 cars the auto industry is putting on the highways this year.

Billions of dollars are being spent to broaden automation's areas of usefulness, yet experts seldom agree on what it is or what it signifies. No development in economic history has generated more discussion or division than this harnessing of electronic brains to mechanical muscles.

The distinctive element in automation is its accent on an unbroken flow of work, as against the stop-and-go methods that stall even the most advanced forms of mass production. "Look, Ma, no hands" is the theme song of the automated factory line. Machines pass parts to one another, give orders to one another, inspect their own product, correct their own mistakes—and the only limit on how much work can be automated is whether an industry has enough customers and enough standardization of product design to justify the cost of automating.

Some analysts consider the whole process a second industrial revolution, with all the potentiality for social upheaval that marked the birth of the factory a century and a half ago. Others insist it is just another step in industry's progress toward greater efficiency, no different in its basic attributes from any of the technological advances that have helped raise American wages, employment totals and living standards and made it possible for us to produce nearly half the world's heavy goods with only 6 per cent of the world's people.

Congressional investigators, puzzled about what action the Government should take, have been told by union leaders that automation threatens mass unemployment and by business executives that it will bring unparalleled prosperity.

Engineers say push-button factories may eventually permit a work schedule in which the week-end will be longer than the week. Educators see all this leisure promoting a scholastic renaissance in which cultural attainments will become the yardstick of social recognition for worker and boss alike. Gloomier observers fear the trend toward "inhuman production" will end by making men obsolete if nuclear warfare does not exterminate them first.

One way to cut through this speculative fog is to visit a plant in which automated and nonautomated operations function side by side. That is the way things run at the Ford engine plant here, where a traditional production line turns out 300 Lincoln and truck engines a day while an automated line puts together eighteen times that many motors for Ford V-8's.

Giant machines capable of turning a 180-pound engine block as if it were made of foam rubber have taken the grunt and groan out of work on the Ford line. An interlocking system of electrical switches and relays, automatic conveyors and air and hydraulic lifts carries the blocks through hundreds of cutting and drilling operations without the touch of a human hand.

The drudgery of hoisting the block onto the machine and pushing it into the precise position required to make the holes come out right is gone. Now the machines do all the sweating. From the time the iron casting is trucked in from the foundry and fed into the first huge broaching machine to have its rough edges shaved off, automated equipment is responsible for transferring it from one machine to another, clamping it into place and giving it a closer going over than any barber gives a man who wants his whiskers off.

Panels of green, red, blue and amber lights—"Christmas trees," the men in the plant call them—show the watching machine attendants what is happening at each step of the production cycle. That is the extent of their physical exertion, except when something needs changing or the machine develops indigestion.

A flash, and a massive cradle spins the casting 45 degrees. Another flash, and a glistening line of manicured metal appears along its side. A third flash, and the block is off toward a battery of drills, so close packed they look like a steel forest but tilted and tuned to function with the precision of a symphony orchestra.

When the rhythm breaks, the lights show where to look for the trouble. The machines do not stop with keeping track of their own health. They also check their product to make sure that the width and depth of each hole is right to the ten-thousandth of an inch. Steel fingers reach down to "feel" for obstructions. If anything is wrong, the gauging device signals the machine to stop until the damage is repaired.

Automatic tool control boards give advance notice that drills and cutting edges are wearing out. If the notice is not heeded in time, they shut off the machine themselves and it stays shut until replacement tools are inserted from a pre-set reserve. Automation even acts as its own janitor. Turnover fixtures upend the ponderous blocks to drain off cutting lubricants and shake out metal chips. The shavings fall through a floor grating onto a belt that carries them to a central bank for scrap iron.

The Ford line is far from the ultimate in automation. At least one-third of the operations, especially those concerned with the intricate task of putting together the spark plugs, valves, manifolds, crankshaft, piston rods, cylinder heads and the dozens of other parts that must be assembled to make the finished engine, are still done in the conventional way. But the contrast with the Lincoln line is unmistakable—and arresting.

The Lincoln engine blocks, twenty pounds heavier than those used by Ford, are lifted onto the cutting machine with an air hoist. Once up, the operator must tug and shoulder them into position. The blocks are pushed along a gravity roller from one machine to another. At each stop the tug and push into place must be repeated. Processes that are done by a single multiple-purpose machine on the Ford line may require as many as seventeen machines on the Lincoln side of the plant.

Where watchfulness is the chief requirement of the men who monitor the Ford robots, brawn remains an indispensable ingredient on the Lincoln line. The slower production tempo enables the Lincoln workers to tend two or three machines at the same time, but even with that spread the manpower needs are perceptibly heavier than in the automated sections of the Ford line. There whole departments, covering an acre of floor space, operate with only a handful of button-pushers, tool-setters and maintenance men.

The contrast becomes even more eerie when one wanders

from this deserted atmosphere into the hurly-burly of the final assembly operations. With an engine coming off the Ford final lines every twelve seconds, workers are jammed so tightly that they bump elbows or shoulders as they perform their rigidly routinized tasks in the manner made famous by Charlie Chaplin in "Modern Times" in the Thirties.

But even in this most complex step of the production process, automated equipment has shifted the heavy labor from man to machines. Mechanical arms carry the engine from station to station; finger-tip controls permit the worker to rotate the nearly complete power plant to the angle that makes it easiest to insert each part; electric switch trippers cause the conveyors to bypass a busy station and come back automatically when the station is free.

Sometimes automation is too smart for its own good. That is what happened to a little "qualifying" machine that used to police the gateway to the engine block line. Before the castings entered the first broach, they had to pass examination by the little machine. If a block were too big or too irregular, a warning light blinked on and the block was pulled out of line.

After a few months of heavy discards, Ford officials decided they would be better off with a less exacting watchman. They found that the operator of the broach could do a more satisfactory job of judging whether a block would go through his machine without jamming. Now the little qualifier sits in dark idleness atop the line, a mute victim of its own efficiency.

"This plant was an engineers' paradise," is the epitaph supplied by K. N. Kreske, the plant manager. "We had to knock out a few machines because they were just too idealistic."

Kreske's comment is not intended as a complaint against automation. He is all for it. A brisk, dark-haired Navy veteran, with the build and energy of a football quarterback, he spent fifteen years with General Motors before he came to Ford in 1951. He has been in charge of production on the automated line since it was installed three years ago.

He feels that the new equipment has relieved the plant's 7,900 workers of the tasks no one ever liked to do. Jobs once shunned because they entailed back-breaking toil now are prized for their ease of accomplishment. At the start some of the old-timers who came over from the closed motor and casting machine shops were baffled by the light panels (some are so tall they resemble an elevator signal board in a Wall Street skyscraper).

To overcome this strangeness, the company sent 300 key tool-setters to school until they learned what the lights meant and what to do when they beamed their message. Now many of the automation machine operators and tool-setters have got the "feel" of their machines so well that they rarely look at the panel. They can sense trouble coming even before the lights record it.

The big emphasis is on avoiding "down time"—periods in which the machine is shut down for repairs or tool changes. When one machine is doing the work normally performed by fifteen, every shutdown is costly. In the old days the rest of the line could run almost unimpaired if one machine was out. Now each delay is cumulative. It piles up all along the line. That is why the automatic tool control boards and the other electrical trouble-spotters are so vital. And why the monitors try to keep ahead of the trouble, instead of being behind it.

The men who play nursemaid to the robots like their jobs. They go home less tired; the pay is better; the work is softer. Few complain of boredom. On the contrary, they speak of their mechanical charges with a pride that is rare among factory workers. But it is a pride overlarded with fear. Many share the worry of their union chiefs that the machines will eventually wipe out their jobs unless the company and the Government do something to prevent freedom from drudgery turning into freedom from pay-checks.

Thus far, unemployment has been no problem. With Ford striving to outpace Chevrolet in their duel for mastery of the low-price auto field, all the workers displaced by automation have been absorbed in other operations. Everyone in the engine plant on both the automated and nonautomated lines is working overtime. Most put in ten hours a day, with an additional eight or nine hours on Saturday.

Two-shift operation is general, with a few bottleneck divisions on a round-the-clock schedule. Weekly pay-checks of $125 to $150 a week before deductions are commonplace. The euphoria of today's well-being dims but does not erase the concern that automation has stirred for the future.

One of the few who manifest no anxiety is Sam Allen, a jovial giant, whose towering frame makes him a fit guardian for the mammoth broaching machine that is the first stop for the engine castings.

"Automation has saved me; it has added fifty years to my life," says Allen. "I am 56 now and I have been at Ford twenty-seven years. If I still had to lug those heavy blocks into position like I used to, I could not last till I was 65. Now I expect to keep working till I am 80."

His only complaint is that he has put on thirty-three pounds since he was assigned to automation. He used to weigh 240 pounds, now he is up to 273.

"I don't do nothing but press those two buttons," he says. "Sometimes I use my thumbs, sometimes I use my wrists and sometimes I lay my whole arm across. The only time I sweat on the job any more is when the sun is 100-and-something outside. When I went down to get a suit made last week, the tailor asked me what happened. I said, 'I like to eat. This fat don't bother me, and I don't bother it.'"

A more sober appraisal comes from Cleveland A. Peck, another veteran of twenty-seven years at Ford, who is vice president of the United Auto Workers' unit at the engine plant. An earnest, articulate man, with a son interning at a New York hospital, Peck recalls going home in the Thirties so tired he would fall asleep in the trolley. With automation, he finds more strain on the mind and little on the back and muscles.

Workers are eager to get on the automated jobs, he reports, not because the pay rates are 12 cents an hour higher but because the jobs are less exhausting and more stimulating. Grievances are far less numerous among the workers on automation than in other sections of the plant. The men get along better with one another and with their supervisors.

The popularity of the automated jobs was made clear to Peck three months ago when the termination of a Ford military tank contract brought several hundred high-seniority men into the engine plant and forced the transfer of men from automated operations into more conventional assignments. "The men who were shifted did not like it at all," he says.

Peck is convinced that automation will make rapid strides in every section of auto manufacture. He hopes it will, provided that effective measures are taken to prevent human wreckage from accompanying its spread. He favors a shorter work week and earlier retirement as safeguards against widespread joblessness. Without such props for mass purchasing power, he fears there will be few customers to buy the products automated factories will be able to turn out in such abundance.

The union's campaign for a guaranteed annual wage resulted in establishment of a supplemental unemployment benefit fund that will provide a six-month cushion against layoffs. Some Ford workers are building up nest eggs of their own just in case automation snuffs out their jobs. Saturnino Puente, who sets drills and machine parts on the engine's overhead valve, buys an $18.75 bond every week. Others take outside work for added financial security.

Now that Ford stock is going on the market, the union has urged the company to renew a previously rejected offer to let its workers have stock at half price. That way a man displaced by a machine would have the consolation of knowing that the machine was making money for him as a shareholder.

In the company's executive offices, no one sees anything but good stemming from automation. Del S. Harder, Ford's executive vice president, who is credited with having coined the word, and D. J. Davis, vice president for manufacturing, emphasize that the high cost of automated equipment and its limited applicability to many fast-changing phases of auto-making are bound to make its effects gradual. Far from destroying jobs, they feel its long-term result will be to cause a shift from menial labor to more highly skilled, better paid, safer and more interesting employment.

Without maximum application of automation and other improvements in technology, the nation's problem may not be too many workers, but too few, the Ford high command says. If industry were obliged to meet today's demand for goods with the machines and methods that were available in the immediate postwar period, the total non-farm labor force would be short by at least 10,000,000 workers, their statistics indicate. And prices would be prohibitively high.

What bothers labor is that, while the assurances of gradualness are being given, all of the automotive Big Three are pressing billion-dollar expansion and modernization drives, with increased automation as a principal target. Plymouth has just unveiled a quarter-mile-long engine assembly machine that cost $2,500,000, cut assembly costs in half and shaved nearly a quarter off the old labor requirements. Similar machines are being built for Ford and General Motors.

Not the least paradoxical element is that, in this dawn of the age of automation, the proudest car in the whole Ford family is the hand-assembled Continental. You can pick one up for a mere $10,000 if you have an automated money press handy.

Perhaps this seeming contradiction offers part of the answer to what automation means. For it suggests once again that, as we find new and better ways to have machines work for us, we develop a greater craving for—and a greater ability to enjoy—the fruits of craftsmanship. We reach out for services of hand and brain to adorn and enrich life.

The visitor to the Rouge plant comes away with a conviction that we are on the threshold of untold abundance. The road is open for a productive system in which machines do the dirty work and all the monotonous repetitive operations that tire the body and dull the mind. Far from being thrown onto the junk heap by the machines' efficiency, man will acquire a new dignity.

The white-collar worker will become as familiar on the production line as in the office. With fewer people needed in farm and factory to satisfy our physical wants, more stress can be put on the sciences, arts and professions. There will be more education and more people engaged in medicine, teaching, city planning, art and other cultural pursuits. Imaginative new skills will be required in distribution to assure an adequate market for the limitless products of industry.

All these things are less than a generation away if we have the will and the vision to grasp them. What is needed is a program of retraining and economic cushions to guarantee that there will be a minimum of human suffering and dislocation on the highway to the new Utopia.

Automation Is Here to Liberate Us

by Eric Hoffer

SOMETHING STRANGE happened to me early in 1964. After a life-time of hardly ever sticking my nose outside the San Francisco waterfront, I found myself running around shooting my mouth off, telling people of an impending crisis, a turning point as fateful as any since the origin of society, and warning them that woe betides a society that reaches a turning point and does not turn. For the first time in my life I became possessed by something.

I am not by nature one of the doom-around-the-corner boys. But this automation thing has hit me between the eyes and I haven't recovered yet. I am convinced that in a matter of decades the people I have lived and worked with all my life will become unneeded and unwanted.

Now, at one point in history God and the priests seemed to become superfluous, yet life went on as before. Then again, the aristocrats became superfluous and hardly anyone noticed their exit. In Russia, where they have capitalism without capitalists, businessmen are superfluous, yet things get done somehow. In

From the *New York Times Magazine,* October 24, 1965, copyright ©
1965 by The New York Times Company.

this country, until yesterday, the intellectuals seemed superfluous, yet we built America. But when the masses become superfluous it means that humanity is superfluous, and this is something that staggers the mind.

Here are some of the statistics. They come neither from labor unions nor from radical or liberal sources, but from the daily press and national magazines. In 1963, Mr. John I. Snyder Jr., one of the foremost manufacturers of automation equipment, told a Congressional committee that automation was eliminating 40,000 jobs a week. Another source put the figure at 50,000.

In other words, already in 1963 automation was eliminating between 2 and 2.5 million jobs a year. At the same time, in 1963, 2.5 million young people entered the labor market. In 1964 it was 2.7 million, in 1965 it will be 3.5 million, and in 1970, 4 million young people will enter the labor market.

We are told that once the economy starts to grow at a satisfactory rate it will absorb most of the unemployed. I think this is a pipe dream. Eighty per cent of the money spent on growth is spent on labor-saving devices. It takes at present a $30,000 increase in the gross national product to create one job. Ten years ago it took $12,000. Ten years from now it may take $75,000. No one expects our economy to grow faster than 5 per cent a year. With a gross national product of $600 billion, 5 per cent comes to $30 billion, and $30 billion creates only one million jobs. Indeed, this is the figure you come across in reports about the number of new jobs the economy is creating. Our economy would have to grow at an astronomical rate to absorb the millions of unemployed.

We have, of course, the official statistics, which, at last report, show the number of unemployed to be about three million. The people who compile these statistics are honest and competent, but they have to follow certain criteria. They can call a person unemployed only when he has done no work whatever, and has "actually sought a job recently." Those who gave up looking for a job because it was no use are not counted among the unemployed. On the other hand, a person is counted as employed if he works only one hour a week.

Recently the California Department of Finance adjusted the Federal statistics for both part-time employment and the Labor Participating Force. (The Labor Participating Force is the labor force expressed in percentage of the total adult civilian population; it had shrunk from 58.5 per cent in 1953 to 55.6 per cent in 1963.) The Department came up with an unemployment rate for California of 11.7 per cent—more than twice the Federal figure. That would mean more than six million unemployed for the whole country.

No matter what the official statistics say, we shall wake up one day to find 20 million or so unemployed in our midst, and a full-blown crisis on our hands.

Now what worries me about the prospective 20 million unemployed is not that they will starve. Poverty can be solved by money, although up to now the poverty already resulting from unemployment by automation has not been tackled. In eastern Kentucky and West Virginia, where the mines have been fully automated, poverty has been allowed to run its course. The automated machines, three stories high, dig coal a thousand feet below the ground, and these post-historic monsters are surrounded by a demoralized population dominated by a feudal ruling class made up of relief officials, mine officials and school superintendents.

No. I do not think that with 20 million unemployed the Appalachian pattern will repeat itself in other parts of the country. I assume that the superfluous population will be given the wherewithal for satisfactory living, even enough to buy things and go fishing.

What worries me is the prospect of a skilled and highly competent population living off the fat of the land without a sense of usefulness and worth. There is nothing more explosive than a skilled population condemned to inaction. Such a population is likely to become a hotbed of extremism and intolerance, and be receptive to any proselytizing ideology, however absurd and vicious, which promises vast action. In pre-Hitlerian Germany a population that knew itself admirably equipped for action was rusting away in idleness and gave its allegiance to a Nazi party which offered unlimited opportunities for that action.

In this country, even the inaction due to retirement often becomes explosive. Southern California, the promised land of retired generals, admirals, business executives, farmers and shopkeepers, is a breeding ground for all sorts of cults, utopias and extremist groups. My feeling is that an energetic, skilled population deprived of its sense of usefulness would be an ideal setup for an American Hitler.

Yet it is part of the fantastic quality of human nature that the thwarted desire for action which may generate extremism and intolerance may also release a flow of creative energies. There are examples from every era illustrating this fact, and none more striking than the conditions which attended the first appearance of written literature in the ancient civilizations.

We are often told that the invention of writing in the Middle East about 3000 B.C. marked an epoch in man's career because it revolutionized the transmission of knowledge and ideas. Actually, for many centuries after its invention, writing was used solely to keep track of the intake and outgo of treasuries and warehouses. The earliest examples we have of writing are invoices and lists of articles.

Writing was invented not to write books but to keep books. The scribe who practiced the craft of writing was a civil servant —a clerk and bookkeeper. Literature was the domain of storytellers and bards who no more thought of writing down their stock in trade than other craftsmen would the secrets of their trade.

Century after century the scribe went on keeping records. He felt snug in his bureaucratic niche, had no grievance and dreamed no dreams. Then, in every civilization, at some point, the scribe makes his appearance as a "writer." When you try to find out what it was that started the scribe "writing," the answer in every case is the same: The scribe began to write when he became unemployed.

Other examples come to mind of the connection between forced inaction and the release of creative energies. Thucydides was a passionate general. He did not want to be a writer; he wanted to command men in battle. But after losing a battle he was exiled, and had to eat his his heart out watching other generals fight the war. So he wrote "The Peloponnesian War," one of the finest histories ever written.

Machiavelli was a born schemer. His ardent desire was to pull strings, negotiate, intrigue, caucus, go on missions. But he lost his job as a diplomat and had to go back to his native village where he spent his days gossiping and playing cards, his evenings writing "The Prince" and "Discourses on Livy."

One more example. During the reign of Louis XIV the French aristocracy produced a crop of remarkable writers: de Retz, Saint Simon, La Rochefoucauld. If you ask why it happened in France and not in other countries, the answer is again—unemployment. While the aristocracies of England, Spain, Italy and Germany were managing affairs, fighting wars, amassing fortunes, and making and unmaking kings, the French aristocrats were taken off their estates, pulled out of the army, and brought to Versailles where all they could do was watch each other and be bored to death.

Enough has been said to show that a loss of a sense of usefulness and a passionate desire for action may release a creative flow in all sorts of people—in sheepherders, farmers, officials, admirals, politicians, aristocrats, and run-of-the-mill scribes.

It goes without saying that in addition to a thwarted desire for action, there must be talent and a degree of expertise. We can, therefore, expect unemployment to release a creative upsurge in the masses only if we assume that the masses are no less endowed with native intelligence than other segments of the population, and that it is possible to bring about a diffusion of expertise in literature, art, science and other fields comparable to the existing wide diffusion of expertise in mechanics and sports.

The cliché that talent is rare is not founded on fact. All that we know is that there are short periods in history when talent springs up all over the landscape, and long periods of mediocrity and inertness. In the small city of Athens within the space of 50 years there sprang up a whole crop of geniuses—Aeschylus, Sophocles, Euripides, Phidias, Pericles, Socrates, Thucydides, Aristophanes. Something similar happened in Florence at the time of the Renaissance; in the Netherlands between 1400 and 1700, during the great period of Dutch-Flemish painting, and in Elizabethan England.

What we know with certitude is not that talent and genius are rare exceptions but that all through history talent and genius have gone to waste on a vast scale. Stalin liquidated the most intelligent, cultivated and gifted segment of the Russian population, yet no one will maintain that Russia is at present less endowed with talent than before the revolution.

The attempt to realize the potentialities of the masses may seem visionary and extravagant, yet it is eminently practical when judged by the criterion of social efficiency. For the efficiency of a society must be gauged not only by how effectively it utilizes its natural resources, but by what it does with its human resources. Indeed, the utilization of natural resources can be deemed efficient only when it serves as a means for the realization of the intellectual, artistic, and manipulative capacities inherent in a population.

It is evident, therefore, that if we are to awaken and cultivate the talents dormant in a whole population, we must change our conceptions of what is efficient, useful, practical and wasteful. The business of a society with an automated economy can no longer be business. Up to now in this country we have been warned not to waste our time, but we are brought up to waste our lives.

Does this mean that we have to eliminate, or radically change, our present free-enterprise system? Not at all. On the contrary, the state of affairs we are striving for might actually give more leeway to the people who operate, and benefit from, the present system. For we shall free them from responsibility for the unneeded and unwanted millions who would remove themselves to a place where they can experiment with a new way of life.

In other words, we recommend here two social systems coexisting side by side, not in competition and strife but in amity and mutuality, and with absolute freedom of movement from one to the other.

Usually, when we try to think of a substitute for our present system, the choices which offer themselves are, singly or in combination: society as a church, society as an army, society as a factory, society as a prison, society as a hospital, and society as a school. For our purpose, the choice must be the last-named —society as a school.

I am not unmindful of the fact that so far, except in the areas of science and philosophy, schools have not been a forcing house of talent. The best of our literature, painting, sculpture, music, etc., has not come out of schools.

It is also true that the most oppressive and ruthless ruling classes in our present world are made up of intellectuals, including a large number of former schoolteachers. This is true of the Communist countries, of the new nations in Asia and Africa, and of the government by professors in Portugal. But we shall have to take the risk, and provide against the evils of rule by schoolmasters.

I would start out with a pilot state made up of a slice of northern California and a slice of southern Oregon, and run by the University of California. I would call it the state of the unemployed, and anyone crossing into it would automatically become a student. The state would be divided into a large number of small school districts, each district charged with the realization and cultivation of its natural and human resources. Production of the necessities of life would be wholly automated, since the main purpose of life would be for people to learn and grow.

I said that the school districts would be small, for I am convinced that the unfolding of human capacities requires a community in which people of different interests, skills and tastes know each other, commune daily with each other, emulate, antagonize, and spur each other.

It must be a society in which people can teach, and learn

from each other; and whether they sympathize, envy, praise or detract, their hearts flutter, their faces flush, and their minds swell with the desire to emulate and surpass.

The absolute freedom of movement from one system to the other and from one district to the other, will result in a continued sorting out of people, so that eventually each system and each district will be operated by its most ardent adherents.

I am convinced that the coexistence of two different social systems in one country would enhance our sense of freedom. For freedom is predicated on the presence of alternatives in the economic, cultural, and political fields. Even in the absence of tyranny, freedom becomes meaningless where there is abject poverty, political inertness, and cultural sameness. And certainly, no alternative can be as productive of a sense of freedom as the alternative of two different social systems.

As to safeguards against tyranny by schoolmasters: Just as the monopoly of power by an aristocracy or a plutocracy may be broken by creating more and more aristocrats or plutocrats, so despotism by intellectuals can be averted by turning everyone into an intellectual. This is what society as a school would automatically do.

Finally, it would be particularly fitting if the new states of the unemployed were to be created in parts of the country that have been depleted and ravaged—where forests have been destroyed, mines worked out, the soil exhausted. The simultaneous reclamation of natural and human resources would add zest and give a higher congruity to the new societies.

To become possessed by something for the first time, and late in life, is an outlandish experience. As the months went by I noticed how myths and legends came floating into my mind. It was some time before I realized that the myths dovetailed into a pattern, that they were telling a coherent story—a version of automation. Here it is.

When God created the world, He immediately automated it, and there was nothing left for Him to do. So in His boredom He began to tinker and experiment. It was in a mood of divine recklessness that God created man. ". . . in the image of God created He him." And it was a foregone conclusion that a creature thus made would try to emulate and surpass his creator. And, indeed, no sooner did God create man than He was filled with misgivings and suspicions. The moment man ate from the tree of knowledge God's worst fears were confirmed. He drove man out of Eden and cursed the ground he would have to till for good measure.

But you do not stop a conspirator from conspiring by exiling him. Though condemned to wrestle with a cursed earth for his bread, and fight off thistles and thorns, man resolved in the depths of his soul to become indeed a creator—to create a man-made world that would straddle and tame God's creation.

Thus all through the millennia of man's existence the vying with God has been a leading motif of his strivings and efforts. Much of the time the motif is drowned by the counterpoint of everyday life, but it is clear and unmistakable in times of great venturesomeness.

In the late Neolithic and Bronze ages "when men began to multiply upon the face of the earth" and in a burst of creativeness invented the wheel, sail, plow, brickmaking, script, calendar, fermentation, irrigation, metallurgy and other momentous devices, they also set out "to build a tower whose top may reach into heaven." They said they were building the tower for the glory of it, "to make us a name," but God knew better. "Behold," He said to His retinue of angels, ". . . this they begin to do, and now nothing will be restrained from them, which they have imagined to do." So He confounded their language and scattered them abroad upon the face of the whole earth. It was only 6,000 years later that the modern Occident picked up where the builders of the Tower of Babel left off.

It was the machine age that really launched the second, or man-made, creation. The machine was man's way of breathing will and thought into inanimate matter. Unfortunately, the second creation did not quite come off. Unlike God, man could not immediately automate the man-made world. He was not inventive enough. Until yesterday, the machine remained a half machine;

it lacked the gears and filaments of will and thought, and man had to use his fellow men as a stop-gap, to yoke men and women and children with iron and steam.

The machine age became a nightmare, an echo of the fearful tale of the Bull of Phalaris. This story tells of an Athenian artist who made a brazen bull for the King of Phalaris. The bull was so lifelike that the artist was seized with a desire to make the bull come alive and bellow like a real bull. So he constructed the throat of the bull in such a way that, when a human being was placed inside the belly and a fire lit underneath, the shrieks and groans of the victim as they came through the specially con-structed throat, sounded like the bellowing of a live bull.

Just so during the past 150 years have millions of human beings been scooped off the land and shoved into the bellies of smoke-belching factories to make the Bull of Phalaris roar, the machine age consuming them as fast as it consumed coal.

Then yesterday, almost unnoticed, the automated machine edged onto the stage. It was born in the laboratories of technical schools where mathematicians and engineers were trying to dupli-cate the human brain. And it was brought into the factory not to cure the disease of work which has tortured humanity for un-told generations, but to eliminate man from the productive process.

Power is always charged with the impulse to eliminate human nature, the human variable, from the equation of action. Dictators do it by terror or by the inculcation of blind faith; the military do it by iron discipline; and the industrial masters think they can do it by automation. But the world has not fallen into the hands of commissars, generals and the National Association of Manufacturers.

There is a change of climate now taking place everywhere which is unfavorable to the exercise of absolute power. Even in totalitarian countries the demands of common folk are be-coming determining factors in economic, social, and political decisions. There is, therefore, a chance that the dénouement of automation will be what we want it to be.

The fact is that the mad rush of the last hundred years has

left us out of breath. We have had no time to swallow our spittle. We know that the automated machine is here to liberate us and show us the way back to Eden; that it will do for us what no revolution, no doctrine, no prayer and no promise could do.

But we do not know that we have arrived. We stand panting, caked with sweat and dust, afraid to realize that the seventh day of the second creation is here, and the ultimate sabbath spread out before us.

Myth of the New Leisure Class

by David Dempsey

LEISURE: *condition of having one's time free from the demands of work or duty.* —AMERICAN COLLEGE DICTIONARY

If we are to take literally the findings of statisticians, ours is a civilization that works less and plays more than any since the Roman Empire declined and fell. No less than sixty million of us belong to the new Leisure Class, a group whose hands have been freed by technology and whose minds are now being liberated by automation. We are men and women, manual workers and white-collared, old and young, all with plenty of free time (theoretically, at least), a steady income and, most important, an abundance of available credit.

On paper, there is no argument about it. Salaried employes and wage-earners work an average of forty hours a week where, as recently as 1929, their fathers worked fifty, and congratulated themselves on not having to work sixty, as *their* fathers did in 1900. (Today's workers also produce six times as much as their grandfathers produced for every hour on the job). What is more, as the work week shrinks, the vacations get longer—a month is no longer unusual for an employe with seniority.

On paper, too, this huge dividend of extra time made possible by modern technology is an unmixed blessing. We spend more money on recreation and amusement than we did a generation ago on food and clothing. Including travel and the do-it-yourself movement, some $34 billion a year finances what is represented to us as a longer and longer pause in the day's occupation.

One student of the subject, for instance, credits the average American with about 3,000 "free" hours a year—spare time, if you will, in which we can cultivate everything from our minds to our gardens. Because of this, we are the envy of the whole world, although, in our more reflective moments, we often wonder why.

Indeed, as someone who has had scarcely a minute to himself in the past twenty years, I am inclined to regard the Age of Leisure as a tantalizing myth. Bemused by statistics, we ignore the facts, and the facts are that the majority of us work harder— at something—than we have ever done before. There are reasons for this, of course, and one of them is that technology, the great liberator, has fooled us. In raising our standards of living, it has added new working hours which must be exchanged for the automatic driers and hardtop convertibles that make up this standard.

If we were willing to live in the style of our grandfathers, even the twenty-hour week would be a snap, but as things stand, industrial inventiveness is the carrot on the stick—creating new products, it also creates the urge to work for them. But this is not all. Our highly organized society, with its increasing "work load" of community responsibilities, eats up time with the rapacity of a wood-burning locomotive consuming fuel. Finally, what free time is left to us is spent not so much in the enjoyment of leisure as in the frantic pursuit of it.

Let us examine these three myth-making factors in somewhat greater detail.

The myth of the declining work week.

In the last seventeen years, from 1940 to 1957, the average number of hours spent on the job by production workers each week has risen rather than declined. For last year, the figure was 40.1 hours per week, whereas in 1940 it was 38.1. Since these averages include seasonal industries such as construction and mining, they do not indicate the extent to which many nonseasonal employes work a forty-two- or forty-three-hour week.

In addition, one in every twenty workers, according to the U.S. Census Bureau, is holding down two jobs, and some—100,-000, to be exact—are managing three. Even among the one-job men, the chief advantage of the "contract week" is that it gives them a chance to work more overtime—the national average is two hours a week.

There are some occupations where forty-five to forty-eight hours a week is the accepted minimum—policemen and nurses, for example, in many communities. In any event, more than 18,-000,000 men and women in the United States do not even pretend to work a forty-hour week—the self employed, household workers, and business men.

Store owners can expect to put in as many as fifty hours and doctors may do even better—or worse, depending upon how you regard work. Farmers still work at least fifty hours a week, despite mechanization. A study by Fortune magazine a few years ago revealed that top corporation executives spend sixty hours a week at their jobs, including the work they take home.

White collar workers fare best, because most of them are limited to a five-day week, with little chance for overtime. But it is among this group that we find a high proportion of working wives. In effect, married women who take employment—and nearly a third of them do—are simply adding a job to one they already have (as housewife), while men whose wives work also take on the extra job of helping to run the house.

Paradoxically, for a "leisure" society, more women work for

a living than ever before. (For that matter, more of everybody works—53.5 per cent at the last census in 1950 compared with 47.3 per cent in 1880.) They have been liberated from their kitchens, by modern labor-saving devices, right into the factories, proving, I suppose, that for some, work is not so bad after all, and that for others it is absolutely necessary to keep up with the high cost of pursuing leisure.

The myth that free time is leisure time.

In view of the foregoing, let us deduct from the 3,000 hours allocated to us by the statisticians, 1,000 hours for overtime, dual employment, and other "hidden" forms of work which do not show up on most reports. (True, some Americans perform no work at all, and may be entitled to unlimited leisure, but we are speaking here of averages.) What happens to the 2,000 hours that are left?

This is precisely the question that baffles the new man of leisure. The answer, I think, is to be found in Rudy Vallee's theme song of some two decades ago, "My time is your time." For it is not until the typical American breadwinner gets home at the end of the day that the pressures of our society really start to close in on him.

A typical week for one middle-class family (and who isn't middle-class anymore?) may go something like this: On Mondays, husband and wife attend an adult class in ceramics at the High School. Bowling on Tuesday (for the man); volunteer work at the hospital on Wednesday (for the woman); square dancing on alternate Thursdays (for both). Friday is an "open night," which means that they may take the children to the movies ("togetherness"), play bridge, collect money for one of the numerous charity drives in the community, or—as they have been promising themselves for some time—read a Great Book.

It can be argued that people don't have to do these things, and there are, of course, in every town a number of hold-outs who have never punched a doorbell in behalf of the Red Cross or eaten supper in a church basement. There are others who spend their evenings in front of the television. But here, too, we are speaking of averages and most families find it less painful to face up to their obligations in the P.T.A., the Cub Scouts and other worthy organizations than to avoid them. Whether for good or bad, we are a nation of joiners and we join, by and large, what we are expected to join.

This might be called the obligatory use of leisure, and although the American male has become a little restive under such a regimen, he gets scant sympathy from his wife.

"You only have to worry about your evenings," she says. "I've got to contend with it all day long."

The pace is grueling. A housewife I know, struggling to keep up with her committee work, her ticket selling and ticket buying, her clubs, telephoning, benefits and church activities, recently threw everything over for a part-time job in a book store. "I just had to have some time to myself," she explained.

Two years ago, The Farm Journal polled its women readers with the question: "What do you do with the time modern home-making equipment frees you for?" The answer indicated that a majority of these women were raising larger families, doing practical nursing, running beauty shops and teaching school. A fair number took up a hobby, but in discouragingly few cases had they acquired any additional leisure.

These women, most of whom do not have access to the more highly organized social life of towns and cities, may not be typical of American women as a whole. They do, however, indicate the dilemma people face when they are suddenly liberated from the established routine.

In most situations, the problem is "solved" by the continually expanding demands of the community, and although this may be a pause in the day's occupation, it is seldom the pause that refreshes. Let us, therefore, in the spirit of local pride and civic endeavor, deduct another 1,000 hours from the 2,000 remaining to us as the heirs of automation. This residue of free time is surely ours to enjoy in the true spirit of leisure, which, in its Latin root (*licere*), means simply "to be permitted."

But is it?

The myth that we get our leisure in play and recreation.

This factor is in some ways the most baffling of all, for it is during these 1,000 hours a year that we play tennis, go to ball games, watch television, barbecue spare ribs and otherwise comport ourselves in the leisure mode. There is scarcely a self-respecting municipality in the country that does not have an organized recreation program, conveniently ordered so that Papa can get in a set of tennis while Junior is playing with the Little League.

Twenty million of us bowl; five million Americans own pleasure boats which twenty-five million of their friends ride in. Ten million homes are tooled up with modern carpentry shops. Hardly a man is now alive, moreover, who does not boast a hobby, or who is not about to take one up.

Here, the statistics are impressive and there is no doubt that our biggest strides toward achieving a leisure society have been made in this direction. Thirty-four billion dollars annually for recreation, travel, and do-it-yourself may not be quite as high as our military budget, but it proves that we rank national fun second only to national defense.

And yet the total effect of this smörgasbord of activities is curiously unsatisfying. For many of us, they are the hardest work we do, adding up, more often than not, to the pursuit of leisure rather than leisure itself. Our week-ends are more strenuous than our working week. Vacations have a way of leaving us exhausted. Competition for our "free time" has become a huge, commercial effort with all the attendant distractions.

"I seldom see a patient who is genuinely suffering from overwork any more," a physician said to me, "but I get plenty of men and women who are driving themselves crazy trying to keep up with the social life and take part in the recreation program that has overrun this town."

This is compulsive, regimented leisure. "People do, not so much what their innermost selves might lead them to do, as what conformity requires in order to rise in the social or economic scale," George Soule, the economist, has written. Hence the frantic urge to "own a boat," to play golf, ski, and to "do" Europe by air (and on credit).

If this is not true leisure, what, then, will the Age of Leisure be like, for although we have not yet entered it, we are without doubt standing nervously on its threshold. Automation today is in its infancy, and the thirty-hour week (and even the twenty) is by no means an improbability. When this happens, there will be a limit to how many committees we can serve on, just as there must be a limit to our tolerance for watching television. The time may not be too far off when at least 1,000 of those mythical hours may really belong to us.

Students of the subject are agreed that the test of leisure is that it should be an end in itself, and that it must represent the individual's own choice of activity. "Leisure isn't just killing time, and it can't be measured by the hours one has off from work. It's a positive period in which people choose what they want to do," writes Prof. Rolf B. Meyersohn, research director of the University of Chicago's Center for the Study of Leisure.

This being the case, how do we get it? Although there is no unanimity of thinking on the question, opinion encompasses the following major points:

(1) Our society must become "leisure directed"—that is, it must learn to relax. As a nation, we have been molded by the pragmatic psychology of the frontier and a Puritan distrust of idleness. When applied to "free time," these once-powerful forces can be self-defeating. We must learn to "play" without feeling guilty about it.

(2) In the words of one expert, our new leisure will be based on a "democracy of interests rather than an anarchy of pursuits." This may involve more and bigger national parks, more art museums, more libraries, more programs for the constructive use of free time, freely chosen. Generally speaking, these activities constitute a nation's "culture"; they are values that can best be shared in leisure activities.

In the past, this culture has been the product of a class which was small in numbers and refined in taste. But in the new mass leisure, as Clifton Fadiman has phrased it, any number can play.

How they play will be the test of viability for our civilization.

(3) We will have more professionally trained leaders in the field of recreation and adult education, and one of their jobs will be to keep us from trying to do everything—in a word, to rationalize the use of leisure. Some day—who knows?—we may even have a Secretary of Leisure in the Cabinet.

(4) No matter how few hours our jobs require of us, we will always "work," but this "work" will be something we want to do. As modern industry continues to relinquish the personal skills that were once so necessary to it—and to us, as craftsmen—we will seek to salvage them elsewhere, at the work bench, the easel or the piano.

Psychiatrists note approvingly that some of our most meaningful types of recreation—gardening, carpentry, fishing, sailing—are really jobs without pay. Indeed, they are among mankind's earliest means of earning a living. The Age of Leisure will see less passive amusement and more direct participation (a trend that is already well on its way).

Whatever the answers, the problem is the same for everyone: how to manage leisure rather than let leisure manage us, the responsibilities of free time being infinitely more difficult than the rigors of an eight-hour day, and the rewards—potentially—being infinitely greater. In the era of Total Automation, leisure rather than work will constitute the dominant values of men's lives.

For some, this may mean the study of philosophy, a chance to become a skin diver, or more time for bird-watching. For others, freed of the compulsive attitude toward the use of free time, it may simply mean that they will no longer feel the need to go fishing on their days off.

The Farm Revolution Picks Up Speed

by Vernon Vine

DOWN SOUTH in the cotton country, out West where they raise sugar beets, and in the dairy lands of the North, farmers are talking machinery.

The news about machinery is exciting. It is a dramatic new chapter in the story of agriculture's mechanical revolution that began when John Deere invented the moldboard plow and Cyrus McCormick perfected the reaper. A new group of machines, just coming into use, or being readied for market, will produce not only bigger farm crops, but a bumper crop of social and economic consequences as well.

Among the most impressive of these machines are the mechanical cotton picker, forage harvesters for the hayfields, sugar-beet, peanut and potato harvesters and corn pickers.

Flame cultivators are important factors in the mechanization of both cotton and sugar beets. So are machines which thin both crops. Barn cleaners, manure loaders and silo unloaders offer farmers relief from some of their most odious and backbreaking chores.

There are even such exotic machines as tree shakers, which

From the *New York Times Magazine,* June 30, 1946, copyright © 1946 by The New York Times Company.

knock nuts off trees, and vacuum cleaners which pick the nuts off the ground (and which will harvest cranberries, or clean out chicken houses as well). There are adaptations of the Army's fog generators, which may make economically practical the use of new, more effective but higher-priced insecticides.

Probably the most significant of these new machines, because of its social and economic implications, is the mechanical cotton picker. This factory-built field hand costs about $5,000, and cuts the cost of growing cotton by $25 a bale when the wage for hand-picking is $1.50 per 100 pounds (considerably less than current wages).

Although the cotton picker was first announced a decade ago, its acceptance was delayed by a number of factors. It was high priced. It gathered leaves as well as cotton, and green bolls as well as ripe ones. Important as these objections were, however, they were not the paramount drawback.

Cotton economy requires that the crop be mechanized in its entirety. So long as cotton growers remain dependent upon hand labor for such jobs as chopping (thinning) and weeding, they might as well keep their sharecropper system, and plow and plant with mule power, and pick by hand also.

But now machines have been devised to chop cotton. Flame weeders have outmoded the hoe. Chemical defoliants, sprayed from airplanes, cause the leaves to drop in the late summer, so the sun can get to the bottom-most bolls, causing them to ripen and open uniformly with the bolls at the top of the plant. Thus, the yield per acre that can be harvested mechanically is increased, and most of the problem of trashy cotton is solved.

Mechanization also has come apace to the hayfield. If your interest in farming is esthetic, rather than economic, you will resent what this means to the rural scene. The haycock is gone, and the hay rack is rapidly going. On some farms, equipped with driers capable of handling freshly cut hay, forage harvesters cut the hay, chop it, and blow it into trailing wagons. The chopped hay is hauled to dehydrators, unloaded mechanically, and the grass is converted into bone-dry feed, all leaves, vitamins and proteins intact.

On other farms, the hay will be mowed as usual (although power-driven mowers do a better job than old friction-drive models). After a few hours' curing, it will be picked up by balers pulled behind tractors, the balers automatically depositing the bales in a wagon behind. Or the hay may be picked up by a field chopper and blown into a wagon. At the barn, curing will be completed in a mow fitted with a huge electric fan, blowing air through a series of ducts.

Sugar-beet growers in the West confidently expect that within five years machines will reduce their labor requirements by 50 per cent. The first problem their engineers went to work on was what to do with the beet seed. It grows in the form of a ball, containing several seed germs. The engineers learned how to split the seed so only one plant would grow where several grew before. They devised a planter which evenly spaces these segments, so that much of the thinning job has been eliminated.

Next came mechanical "blockers," which do for beets what mechanical choppers do for the cotton crop. Weeding now can be done with "sizz-weeders," to give agricultural flame throwers their accepted name. And now, to complete the mechanization of the crop, comes the mechanical harvester. A beet has to be plowed out of the ground. (It is backbreaking to pull up a beet that has not had the soil loosened around it.) It has to have its leafy top sliced off. It has to be pitched onto a wagon or truck. The new harvester does all these operations at once.

Similar machines have been devised for the potato harvest. They dig potatoes, pick them up, shake the loose dirt off, and deposit them in bags. Peanut harvesters, which reduce by seven-eighths the labor required to harvest an acre of peanuts, also have been developed. Peanut shellers, that can do in an hour what a man can do by hand in 300 hours, have been perfected.

Corn pickers are not so new as many other machines, but more of them are coming into use each season, and new models are bigger and more efficient. Some of the current pickers harvest as many as four rows at a time; some are self-propelled, and some have a built-in stalk-shredder which provides automatic control of the corn borer. One casualty of the corn picker is one of the nation's largest sporting events—the national corn husking championship, which used to draw crowds larger than the Army-Navy game.

The self-propelled principle applied to grain combines (no tractor is needed to pull them) has resulted in strange-looking machines which proved important factors in getting the bumper wheat crops into the bin in the last two labor-short harvests.

On the dairy farm, barn gutters now can be cleaned mechanically by an arrangement of electric motor, chain, sprockets, and angle irons which deposit the contents of the gutter into a manure spreader without a hand ever being put to a shovel. Tractor-mounted manure forks take the load off the cattle-feeder's back when the time comes for him to clean out his feed lot.

Mechanical chicken pickers—a blur of rapidly revolving rubber "fingers"—strip the feathers from broilers, capons and turkeys in a matter of seconds. New tillage tools, especially designed for the smaller farm, in one operation plow, disk and harrow, while propelling themselves.

Electricity, in an estimated 200 farm uses, already serves 2,-500,000 farms, and may come to as many more within the next five years. The final touch, however, has been supplied by a Wisconsin manufacturer, who has designed a glass-lined silo, complete with power unloading device.

There can be no mistaking the intention of farmers to mechanize as rapidly as they can. (Because of labor disputes, the outlook for machinery production is poorer this year than last.) They have the money. They also are thoroughly fed up with their labor problem—high wages, inexperience, unreliability, and in many cases no help of any kind at any price. They know that economics favors more machinery. A recent survey indicated machinery costs are up only 14 per cent from prewar; labor costs, 180 per cent.

The effect of this expansion of farm mechanization is bound to be far-reaching. The result, in the case of cotton, may be nearly as significant as the invention of the cotton gin. The price of the cotton picker puts it far beyond the reach of small growers. They are not going to buy pickers. Instead, they eventually will go out of the cotton business, and some out of farming.

The cotton picker also spells the end of the sharecropper and his mule. And as the mule gives way to the tractor, millions of acres that have grown corn to feed him may now be seeded down to pasture and hay to feed still more cows.

As in the other mechanized areas of the nation, the coming of the tractor in the South means fewer farms and larger ones; fewer towns, and larger ones; fewer schools, and larger ones; fewer farm families, and smaller ones. The rural South faces the prospect and problem of catching up with the twentieth century fifty years after it began.

The effects of mechanization on the economics of cotton seem fairly predictable. Shortly before he resigned as Secretary of Agriculture, Claude Wickard proposed a five-year plan for subsidizing the mechanization of cotton production on land best suited to it, and the subsidizing of other types of farming on land now in cotton but not well adapted to mechanical production methods. At the end of the five-year period, Mr. Wickard predicted, the combination of mechanization and the concentration of the crop on the best-suited land not only would enable cotton growers to get along without subsidies but also would make it possible for them to compete on the world markets with foreign cotton, and on the domestic market with synthetic fibers.

Mr. Wickard's plan apparently was buried when he became Rural Electrification Administrator. His successor, Clinton P. Anderson, gloomily recognizes the arrival of mechanization by predicting that it will probably depress prices, although he believes it will benefit agriculture in the long run.

Unless the Wickard plan, or a similar one, is adopted, the adjustment period is bound to be painful. Small farmers, manfully but vainly trying to pit their mules and hands against tractors and machines, will cry out for subsidies and more subsidies, each one benefiting the mechanized producer more than the mule-power operator, and thus only further widening the gap between them.

The impact of machinery on the sugar-beet crop will be less

severe because it will involve fewer persons, and because the West's agriculture is more diversified than is that of the Cotton Belt. One result will be less use of transient labor in the sugar-beet producing States. This may disrupt a large share of the West's migratory labor economy, which functions as smoothly as it does because it is based on a succession of crops which provide the workers with a "circuit" of seasonal jobs through the coastal and mountain States.

The political ramifications of beet mechanization are interesting to contemplate. The domestic sugar industry has long urged a policy of continental self-sufficiency for sugar, but they have had to combat an opposition which made the most of the industry's high labor requirements and consequent high-cost operation. If the new machines give the beet-sugar lobbyists effective answers to these criticisms, they may also give our statesmen new headaches as they wrestle with the future of our commercial relations with Cuba and the Philippines.

What the results of new machinery in other branches of farming will be is not quite so apparent, because only some, and not all, phases of production are affected. This much, however, is clear: These new machines mean less drudgery for farmers and their families; better incomes; more leisure, and more widespread enjoyment of modern amenities.

For the nation as a whole it brings closer the day when we must face up to some fundamental problems. As farmers increase their productive capacity through mechanization it becomes more necessary than ever that we develop a system of distribution so the output of our farms and ranches can be consumed at prices fair both to producers and purchasers. Unless the solution to this problem can be found within the framework of a free economy, we face the choice between more Government control and agricultural regimentation than we ever have had, or chaotic surpluses.

No less significant is the effect of farm mechanization on our population pattern. For years metropolitan areas have not produced enough children to maintain their populations. Our cities have grown because our farms and small towns have exported to them their "surplus" young people. Dwindling numbers of farmers, plus the depressing effect of mechanization on the farm birth rate, mean that rural America will fall increasingly short of meeting our metropolitan deficits.

Meanwhile, farmers are not worrying about such abstruse problems. For the rest of the world the air age, or the electronic age, may just be dawning. But on the farm front this is the beginning of the era of the mechanical cotton picker and the automatic barn cleaner. It is the electrical epoch, the self-propelled century. It may conceivably be the time in which the farmer's day can be cut to twelve hours.

The dawn of this new day already is coming up like thunder— the thunder of the exhausts of myriad new and potentially terrifying machines. It will be a bright, streamlined day. There will be few of the nostalgic reminders of the kind of farming dear to the sentimentalists of the old school.

But the modernists, who find beauty in stark functionalism, have before them a rich experience reporting and depicting an agriculture in which the horse is iron and the old familiar odors have been supplanted by the sweet fumes of a Diesel engine.

Understanding McLuhan (In Part)

by Richard Kostelanetz

TORONTO

MARSHALL MCLUHAN, one of the most acclaimed, most controversial and certainly most talked-about of contemporary intellectuals, displays little of the stuff of which prophets are made.
Tall, thin, middle-aged and graying, he has a face of such meager
individual character that it is difficult to remember exactly what
he looks like; different photographs of him rarely seem to capture
the same man.

By trade, he is a professor of English at St. Michael's College,
the Roman Catholic unit of the University of Toronto. Except
for a seminar called "Communication," the courses he teaches
are the standard fare of Mod. Lit. and Crit., and around the
university he has hardly been a celebrity. One young woman now
in Toronto publishing remembers that, a decade ago, "McLuhan
was a bit of a campus joke." Even now, only a few of his
graduate students seem familiar with his studies of the impact of
communications media on civilization—those famous books that
have excited so many outside Toronto.

From the *New York Times Magazine,* January 29, 1967, copyright ©
1967 by The New York Times Company.

McLuhan's two major works, "The Gutenberg Galaxy" (1962) and "Understanding Media" (1964), have won an astonishing variety of admirers. General Electric, I.B.M. and Bell Telephone have all had him address their top executives; so have the publishers of America's largest magazines. The composer John Cage made a pilgrimage to Toronto especially to pay homage to McLuhan, and the critic Susan Sontag has praised his "grasp on the texture of contemporary reality."

He has a number of eminent and vehement detractors, too. The critic Dwight Macdonald calls McLuhan's books "impure nonsense, nonsense adulterated by sense." Leslie Fiedler wrote in Partisan Review: "Marshall McLuhan . . . continually risks sounding like the body-fluids man in 'Doctor Strangelove.' "

Still the McLuhan movement rolls on. Now he has been appointed to the Albert Schweitzer Chair in the Humanities at Fordham University, effective next September. (The post, which pays $100,000 a year for salary and research expenses, is one of 10 named for Schweitzer and Albert Einstein, underwritten by New York State. Other Schweitzer Professors include Arthur Schlesinger Jr. at City University and Conor Cruise O'Brien at N.Y.U.)

What makes McLuhan's success so surprising is that his books contain little of the slick style of which popular sociology is usually made. As anyone who opens the covers immediately discovers, "Media" and "Galaxy" are horrendously difficult to read —clumsily written, frequently contradictory, oddly organized, and overlaid with their author's singular jargon. Try this sample from "Understanding Media." Good luck.

The movie, by sheer speeding up the mechanical, carried us from the world of sequence and connections into the world of creative configuration and structure. The message of the movie medium is that of transition from lineal connections to configurations. It is the transition that produced the now quite correct observation: "If it works, it's obsolete." When electric speed further takes over from mechanical movie sequences, then the lines of force in structures and in media become loud and clear. We return to the inclusive form of the icon.

Everything McLuhan writes is originally dictated, either to his secretary or to his wife, and he is reluctant to rewrite, because, he explains, "I tend to add, and the whole thing gets out of hand." Moreover, some of his insights are so original that they evade immediate understanding; other paragraphs may forever evade explication. "Most clear writing is a sign that there is no exploration going on," he rationalizes. "Clear prose indicates the absence of thought."

The basic themes in these books seem difficult at first, because the concepts are as unfamiliar as the language, but on second (or maybe third) thought, the ideas are really quite simple. In looking at history, McLuhan espouses a position one can only call "technological determinism." That is, whereas Karl Marx, an economic determinist, believed that the economic organization of a society shapes every important aspect of its life, McLuhan believes that crucial technological inventions are the primary influence. McLuhan admires the work of the historian Lynn White Jr., who wrote in "Medieval Technology and Social Change" (1962) that the three inventions of the stirrup, the nailed horseshoe and the horse collar created the Middle Ages. With the stirrup, a soldier could carry armor and mount a charger; and the horseshoe and the harness brought more efficient tilling of the land, which shaped the feudal system of agriculture, which, in turn, paid for the soldier's armor.

Pursuing this insight into technology's importance, McLuhan develops a narrower scheme. He maintains that a major shift in society's predominant technology of communication is the crucially determining force behind social changes, initiating great transformations not only in social organization but human sensibilities. He suggests in "The Gutenberg Galaxy" that the invention of movable type shaped the culture of Western Europe from 1500 to 1900. The mass production of printed materials encouraged nationalism by allowing more rapid and wider spread of information than permitted by hand-written messages. The linear forms of print influenced music to repudiate the structure of repetition, as in Gregorian chants, for that of linear development, as in a symphony. Also, print reshaped the sensibility of Western man, for whereas he once saw experience as individual segments, as a collection of separate entities, man in the Renaissance saw life as he saw print—as a continuity, often with causal relationships. Print even made Protestantism possible, because the printed book, by enabling people to think alone, encouraged individual revelation. Finally: "All forms of mechanization emerge from movable type, for type is the prototype of all machines."

In "Understanding Media," McLuhan suggests that electronic modes of communication—telegraph, radio, television, movies, telephones, computers—are similarly reshaping civilization in the 20th century. Whereas print-age man saw one thing at a time in consecutive sequence—like a line of type—contemporary man experiences numerous forces of communication simultaneously, often through more than one of his senses. Contrast, for example, the way most of us read a book with how we look at a newspaper. With the latter, we do not start one story, read it through and then start another. Rather, we shift our eyes across the pages, assimilating a discontinuous collection of headlines, sub-headlines, lead paragraphs, photographs and advertisements. "People don't actually read newspapers," McLuhan says; "they get into them every morning like a hot bath."

Moreover, the electronic media initiate sweeping changes in the distribution of sensory awareness—in what McLuhan calls the "sensory ratios." A painting or a book strikes us through only one sense, the visual; motion pictures and television hit us not only visually but also aurally. The new media envelop us, asking us to participate. McLuhan believes that such a multisensory existence is bringing a return to the primitive man's emphasis upon the sense of touch, which he considers the primary sense, "because it consists of a meeting of the senses." Politically, he sees the new media as transforming the world into "a global village," where all ends of the earth are in immediate touch with one another, as well as fostering a "retribalization" of human life. "Any highway eatery with its TV set, newspaper and magazine," he writes, "is as cosmopolitan as New York or Paris."

In his over-all view of human history, McLuhan posits four

great stages: (1) Totally oral, preliterate tribalism. (2) The codification by script that arose after Homer in ancient Greece and lasted 2,000 years. (3) The age of print, roughly from 1500 to 1900. (4) The age of electronic media, from before 1900 to the present. Underpinning this classification is his thesis that "societies have been shaped more by the nature of the media by which men communicate than by the content of the communication."

This approach to the question of human development, it should be pointed out, is not wholly original. McLuhan is modest enough to note his indebtedness to such works as E. H. Gombrich's "Art and Illusion" (1960), H. A. Innis's "The Bias of Communication" (1951, recently reissued with an introduction by McLuhan), Siegfried Giedion's "Mechanization Takes Command" (1948), H. J. Chaytor's "From Script to Print" (1945) and Lewis Mumford's "Technics and Civilization" (1934).

McLuhan's discussions of the individual media move far beyond the trade talk of communications professionals (he dismisses Gen. David Sarnoff, the board chairman of R.C.A., as "the voice of the current somnambulism"). Serious critics of the new media usually complain about their content, arguing, for example, that if television had more intelligent treatments of more intelligent subjects, its contribution to culture would be greater. McLuhan proposes that, instead, we think more about the character and form of the new media. His most famous epigram— "The medium is the message"—means several things.

The phrase first suggests that each medium develops an audience of people whose love for that medium is greater than their concern for its content. That is, the TV medium itself becomes the prime interest in watching television; just as some people like to read for the joy of experiencing print, and more find great pleasure in talking to just anybody on the telephone, so others like television for the mixture of kinetic screen and relevant sound. Second, the "message" of a medium is the impact of its forms upon society. The "message" of print was all the aspects of Western culture that print influenced. "The message of the movie medium is that of transition from linear connections to configurations." Third, the aphorism suggests that the medium itself—its form—shapes its limitations and possibilities for the communication of content. One medium is better than another at evoking a certain experience. American football, for example, is better on television than on radio or in a newspaper column; a bad football game on television is more interesting than a great game on radio. Most Congressional hearings, in contrast, are less boring in the newspaper than on television. Each medium seems to possess a hidden taste mechanism that encourages some styles and rejects others.

To define this mechanism, McLuhan has devised the categories of "hot" and "cool" to describe simultaneously the composition of a communications instrument or a communicated experience, and its interaction with human attention. A "hot" medium or experience has a "high definition" or a highly individualized character as well as a considerable amount of detailed information. "Cool" is "low" in definition and information; it requires that the audience participate to complete the experience. McLuhan's own examples clarify the distinction: "A cartoon is 'low' definition, simply because very little visual information is provided." Radio is usually a hot medium; print, photography, film and paintings essentially are hot media. "Any hot medium allows of less participating than a cool one, as a lecture makes for less participation than a seminar, and a book for less than a dialogue."

The terms "hot" and "cool" he also applies to experiences and people, and, pursuing his distinction, he suggests that while a hot medium favors a performer of a strongly individualized presence, cool media prefer more nonchalant, "cooler" people. Whereas the radio medium needs a voice of a highly idiosyncratic quality that is instantly recognizable—think of Westbrook Van Voorhees, Jean Shepherd, Fanny Brice—television favors people of a definition so low they appear positively ordinary. With these terms, one can then explain all sorts of phenomena previously inscrutable—such as why bland personalities (Ed Sullivan, Jack Paar) are so successful on television.

"It was no accident that Senator McCarthy lasted such a very short time when he switched to TV," McLuhan says. "TV is a cool medium. It rejects hot figures and hot issues and people from the hot press media. Had TV occurred on a large scale during Hitler's reign he would have vanished quickly." As for the 1960 Presidential debates, McLuhan explains that whereas Richard Nixon, essentially a hot person, was superior on radio, John F. Kennedy was the more appealing television personality. (It follows that someone with as low a definition as Dwight Eisenhower would have been more successful than either.)

The ideas are not as neatly presented as this summary might suggest, for McLuhan believes more in probing and exploring—"making discoveries"—than in offering final definitions. For this reason, he will rarely defend any of his statements as absolute truths, although he will explain how he developed them. Some perceptions are considerably more tenable than others—indeed, some are patently ridiculous—and all his original propositions are arguable, so his books require the participation of each reader to separate what is wheat to him from the chaff. In McLuhanese, they offer a cool experience in a hot medium.

A typical reader's scorecard for "Media" might show that about one-half is brilliant insight; one-fourth, suggestive hypotheses; one-fourth, nonsense. Given the book's purpose and originality, these are hardly bad percentages. "If a few details here and there are wacky," McLuhan says, "it doesn't matter a hoot."

McLuhan eschews the traditional English professor's expository style—introduction, development, elaboration and conclusion. Instead, his books imitate the segmented structure of the modern media. He makes a series of direct statements. None of them becomes a thesis but all of them approach the same phenomenon from different angles. This means that one should not necessarily read his books from start to finish—the archaic habit of print-man.

The real introduction to "The Gutenberg Galaxy" is the final chapter, called "The Galaxy Reconfigured"; even McLuhan advises his readers to start there. With "Media," the introduction and the first two chapters form the best starting point; thereafter, the reader is pretty much free to wander as he wishes. "One can stop anywhere after the first few sentences and have the full message, if one is prepared to 'dig' it," McLuhan once wrote of non-Western scriptural literature; the remark is applicable to his own books.

Similarly, McLuhan does not believe that his works have only one final meaning. "My book," he says, "is not a package but part of the dialogue, part of the conversation." (Indeed, he evaluates other books less by how definitively they treat their subject—the academic standard—than by how much thought they stimulate. Thus, a book may be wrong but still great. By his own standards, "Media" is, needless to say, a masterpiece.)

Underlying McLuhan's ideas is the question of whether technology is beneficial to man. Thinkers such as the British critic F. R. Leavis have argued, on the one hand, that technology stifles the blood of life by dehumanizing the spirit and cutting existence off from nature; more materialist thinkers, on the other hand, defend the machine for easing man's burdens. McLuhan recognizes that electronic modes of communication represent, in the subtitle of "Media," "extensions of man." Whereas the telephone is an extension of the ear (and voice), so television extends our eyes and ears. That is, our eyes and ears attended John Kennedy's funeral, but our bodies stayed at home. As extensions, the new media offer both possibility and threat, for while they lengthen man's reach into his existence, they can also extend society's reach into him, for both exploitation and control.

To prevent this latter possibility, McLuhan insists that every man should know as much about the media as possible. "By knowing how technology shapes our environment, we can transcend its absolutely determining power," he says. "Actually, rather than a 'technological determinist,' it would be more accurate to say, as regards the future, that I am an 'organic autonomist.' My entire concern is to overcome the determinism that results from the determination of people to ignore what is going on. Far from regarding technological change as inevitable, I insist that if we understand its components we can turn it off any time we choose. Short of turning it off, there are lots of moderate controls conceivable." In brief, in stressing the importance of knowledge, McLuhan is a humanist.

McLuhan advocates radical changes in education, because he believes that a contemporary man is not fully "literate" if reading is his sole pleasure: "You must be literate in umpteen media to be really 'literate' nowadays." Education, he suggests, should abandon its commitment to print—merely a focusing of the visual sense—to cultivate the "total sensorium" of man—to teach us how to use all five cylinders, rather than only one. "Postliterate does not mean illiterate," writes the Rev. John Culkin, S.J., director of the Communications Center at Fordham and a veteran propagator of McLuhan's ideas about multimedia education. "It rather describes the new social environment within which print will interact with a great variety of communications media."

Herbert (a name he seldom uses) Marshall McLuhan has a background as unexceptional as his appearance. He was born in Western Canada—Edmonton, Alberta—July 21, 1911, the son of mixed Protestant (Baptist and Methodist) parents. "Both agreed to go to all the available churches and services, and they spent much of their time in the Christian Science area," he recalls. His father was a real-estate and insurance salesman who, McLuhan remembers, "enjoyed talking with people more than pursuing his business." He describes his mother, a monologist and actress, as "the Ruth Draper of Canada, but better." His brother is now an Episcopal minister in California.

After taking his B.A. and M.A. at the University of Manitoba, McLuhan followed the route of many academically ambitious young Canadians to England, where he attended Cambridge for two years. There, he remembers, the lectures of I. A. Richards and F. R. Leavis stimulated his initial interest in studying popular culture. Returning home in 1936, he took a job at the University of Wisconsin. The following year he entered the Catholic Church, and ever since, he has taught only at Catholic institutions—at St. Louis from 1937 to 1944, at Assumption in Canada from 1944 to 1946 and at St. Michael's College, a Basilian (C.S.B.) establishment, since 1946.

His field was originally medieval and Renaissance literature, and in 1942 he completed his Cambridge Ph.D. thesis on the rhetoric of Thomas Nashe, the Elizabethan writer. As a young scholar, he began his writing career, as every professor should, by contributing articles to the professional journals, and to this day, academic circles know him as the editor of a popular paperback textbook of Tennyson's poems. Moreover, his critical essays on writers as various as Gerard Manley Hopkins, John Dos Passos and Samuel Taylor Coleridge are frequently anthologized.

By the middle forties, he was contributing more personal and eccentric articles on more general subjects to several little magazines; before long, his pieces had such outrageous titles as "The Psychopathology of Time and Life." By the time his first book, "The Mechanical Bride," appeared in 1951, McLuhan had developed his characteristic intellectual style—the capacity to offer an endless stream of radical and challenging ideas.

Although sparsely reviewed and quickly remaindered, that book has come to seem, in retrospect, the first serious attempt to inspect precisely what effects mass culture had upon people and to discover what similarities existed between mass culture and élite art. Copies are so scarce that they now often bring as much as $40 secondhand. McLuhan had the foresight and self-confidence to purchase a thousand copies at remainder prices; he still gives them to friends, as well as selling them to strangers (at far below the going price). The bottom will soon drop out of the "Mechanical Bride" market, however, for Beacon Press plans to reissue it in paperback and Vanguard, its original publisher, in hardcover.

In 1953, the year after he became a full professor at St. Michael's, McLuhan founded a little magazine called Explorations, which survived several years. Along with a coeditor, the anthropologist Edmund S. Carpenter, McLuhan collected some of the best material in a paperback called "Explorations in Communication" (1960), which is perhaps the ideal introduction to his special concerns and ideas.

Though McLuhan remains a Canadian citizen, he became, in 1959, director of the Media Project of the National Association of Educational Broadcasters and the United States Office of Education. Out of that experience came a report which, in effect, was the first draft of "Understanding Media." Then, in 1963, the

University of Toronto appointed McLuhan to head a newly formed Center for Culture and Technology "to study the psychic and social consequences of technology and the media."

A visitor expects the Center, so boldly announced on its letterhead, to be a sleek building with a corps of secretaries between the corridor and the thinkers. In fact, the Center is more a committee than an institution. It exists, for the present, only in McLuhan's cluttered office.

Bookcases cover the walls, with battered old editions of the English classics on the top shelves and a varied assortment of newer books on Western civilization on the more accessible shelves —6,000 to 7,000 volumes in all. More books and papers cover several large tables. Buried in a corner is a ratty metal-frame chaise longue, more suited to a porch than an office, with a thin, lumpy green mattress haphazardly draped across it.

In temperament, the Center's head is more passive than active. He often loses things and forgets deadlines. The one singular feature of his indefinite face is his mouth. Only a sliver of his lips is visible from the front, but from the side his lips appear so thick that his slightly open mouth resembles that of a flounder. His only visible nervous habits are tendencies to pucker his mouth and push his chin down toward his neck before he speaks, to twirl his glasses around his fingers when he lectures and to rub his fingers down the palms of his hands whenever he says "tactility."

The professor is a conscientious family man. He met his wife, Corrine, a tall and elegant Texan, in Los Angeles, where he was doing research at the Huntington Library and she was studying at the Pasadena Playhouse. Married in 1939, they have six children: Eric, 25; Mary and Theresa, 21-year-old twins; Stephanie, 19; Elizabeth, 16, and Michael, 14, and the girls confirm their father's boasts: "All my daughters are beautiful." Every Sunday he leads his brood to mass.

They live in a three-story house with a narrow front and a small lawn punctuated by a skinny driveway leading to a garage in the back. The interior is modest, except for an excessive number of books, both shelved and sprawled. McLuhan likes to read in a reclining position, so across the top of the living-room couch, propped against the wall, are 20 or so fat scholarly works; interspersed among them are a few mysteries—his favorite light reading. He rarely goes to the movies or watches television; most of his own cultural intake comes via print and conversation. Talking seems his favorite recreation.

McLuhan seems pretty much like any other small-city professor until he begins to speak. His lectures and conversation are a singular mixture of original assertions, imaginative comparisons, heady abstractions and fantastically comprehensive generalizations, and no sooner has he stunned his listeners with one extraordinary thought than he hits them with another. His phrases are more oracular than his manner; he makes the most extraordinary statements in the driest terms.

In his graduate seminar, he asks: "What is the future of old age?" The students look bewildered. "Why," he replies to his own question, "exploration and discovery." Nearly everything he says *sounds* important. Before long, he has characterized the Batman TV show as "simply an exploitation of nostalgia which I predicted years ago." The 25 or so students still look befuddled and dazed; hardly anyone talks but McLuhan. "The criminal, like the artist, is a social explorer," he goes on. "Bad news reveals the character of change; good news does not." No one asks him to be more definite, because his talk intimidates his listeners.

He seems enormously opinionated; in fact, he conjures insights. His method demands a memory as prodigious as his curiosity. He often elevates an analogy into a grandiose generalization, and he likes to make his points with puns: "When a thing is current, it creates currency." His critics ridicule him as a communications expert who cannot successfully communicate; but too many of his listeners, say his admirers, suffer from closed minds.

The major incongruity is that a man so intellectually adventurous should lead such a conservative life; the egocentric and passionately prophetic qualities of his books contrast with the personal modesty and pervasive confidence of a secure Catholic. What explains the paradox is that "Marshall McLuhan," the thinker, is different from "H. M. McLuhan," the man. The one writes books and delivers lectures; the other teaches school, heads a family and lists himself in the phone book. It was probably H. M. who made that often-quoted remark about Marshall's theories: "I don't pretend to understand them. After all, my stuff is very difficult."

And the private H. M. will say this about the technologies his public self has so brilliantly explored: "I wish none of these had ever happened. They impress me as nothing but a disaster. They are for dissatisfied people. Why is man so unhappy he wants to change his world? I would never attempt to improve an environment—my personal preference, I suppose, would be a preliterate milieu, but I want to study change to gain power over it."

His books, he adds, are just "probes"—that is, he does not "believe" in his work as he believes in Catholicism. The latter is faith; the books are just thoughts. "You know the faith differently from the way you 'understand' my books."

When asked why he creates books rather than films, a medium that might be more appropriate to his ideas, McLuhan replies: "Print is the medium I trained myself to handle." So, all the recent acclaim has transformed McLuhan into a bookmaking machine. Late this year, we shall have "Culture Is Our Business," which he describes as a sequel to "The Mechanical Bride." Perhaps reflecting his own idea that future art will be, like medieval art, corporate in authorship, McLuhan is producing several more books in dialogue with others. With Wilfred Watson, a former student who is now an English professor at the University of Alberta, he is completing a history of stylistic change, "From Cliché to Archetype." With Harley W. Parker, head of design at the Royal Ontario Museum, he has just finished "Space in Poetry and Painting," a critical and comparative survey of 35 pairs of poems and pictures from primitive times to the present.

In tandem with William Jovanovich, the president of Harcourt, Brace and World, McLuhan is writing "The Future of the Book," a study of the impact of xerography, and along with the management consultant Ralph Baldwin he is investigating the future of business in "Report to Management." As if that were not enough, he joined with the book designer Quentin Fiore to compile "The Medium Is the Massage," an illustrated introduction to McLuhanism that will be out this spring; the two are doing another book on the effect of automation. Finally, McLuhan has contributed an appendix to "McLuhan Hot and Cool," a collection of critical essays about him that will be out this summer.

On another front, McLuhan and Prof. Richard J. Schoeck, head of the English Department at St. Michael's, have recently produced two imaginative textbooks, "The Voices of Literature," for use in Canadian high schools. And with Professor Schoeck and Ernest J. Sirluck, dean of the graduate school at the University of Toronto, McLuhan oversees a series of anthologies of criticism being published jointly by the Toronto and Chicago University Presses. Obviously, despite the bait from the worlds of media and advertising, McLuhan is keeping at least one foot planted in academia. Only this past December, he addressed the annual meeting of the Modern Language Association on the confrontation of differing sensory modes in 19th-century poetry.

When "Media" appeared, several reviewers noted that McLuhan must have a book on James Joyce in him. That task he passed on to his son Eric, who is writing a prodigious critical study of "Finnegans Wake." Among McLuhan's greatest desires is establishing the Center for Culture and Technology in its own building, with sufficient funds to support a reference library of the sensory experience of man. That is, he envisions methods of measuring all the "sensory modalities" (systems of sensory organization) in all cultures, and of recording this knowledge on coded tapes in the Center. Assistant Professor of Design Allen Bernholtz, one of McLuhan's colleagues, foresees a machine that will, following taped instructions, artificially create a sensory environment exactly similar to that of any other culture; once the subject stepped into its capsule, the machine could be programed to simulate what and how, say, a Tahitian hears, feels, sees, smells and tastes. "It will literally put you in the other guy's shoes," Bernholtz concludes. So far, the projected Center has not received anywhere near the $5-million backing it needs to begin.

Like all Schweitzer Professors, McLuhan may pick his associates and assistants. His entourage will include Professor Carpenter, with whom he coedited Explorations; Harley Parker and Father Culkin. In addition to teaching one course and directing a research project, McLuhan and his associates plan to conduct numerous dialogues and to publish a son of Explorations. "Once you get a lot of talk going," he said recently, "you have to start a magazine." Because he believes that "I can better observe America from up here," he had rejected previous lucrative offers that involved forsaking Toronto, and as Schweitzer professorships are formally extended for only one year (although they are renewable), McLuhan will officially take only a sabbatical leave from St. Michael's.

McLuhan has always been essentially a professor living in an academic community, a father in close touch with his large family and a teacher who also writes and lectures. When some V.I.P.'s invited him to New York a year ago, he kept them waiting while he graded papers. Although he does not run away from all the reporters and visitors, he does little to attract publicity. His passion is the dialogue; if the visitor can participate in the conversation, he may be lucky enough, as this writer was, to help McLuhan write (that is, dictate) a chapter of a book.

"Most people," McLuhan once remarked, "are alive in an earlier time, but you must be alive in our own time. The artist," he added, "is the man in any field, scientific or humanistic, who grasps the implications of his actions and of new knowledge in his own time. He is the man of integral awareness."

Although his intention was otherwise, McLuhan was describing himself—the specialist in general knowledge. Who would dare surmise what thoughts, what perceptions, what grand schemes he will offer next?

McLuhan Weighs Aimless Violence

by John Leo

MARSHALL MCLUHAN'S next book will argue that every new technology necessitates a new war.

In "War and Peace in the Global Village," due in September simultaneously from Bantam and McGraw-Hill, the controversial professor applies his theories on media and technology to the problem of violence.

He has concluded, he said in an interview, that violence is an involuntary quest for identity, and that every new technology sets off this quest by threatening the old personal identity.

"Violence is directed toward image-making, not goals," he said. "The Columbia students have no goals, neither do the Negroes. As long as we provide them with new technology, they must struggle for a new image."

Professor McLuhan, who is teaching at Fordham University this year, is director of the University of Toronto's Center for Culture and Technology and author of "Understanding Media" and "The Gutenberg Galaxy."

The new technology, he said, has provided the whole society

with free or relatively inexpensive "software"—telephone, movies, telegraph and broadcasting—that amounts to a form of communism.

Discrepancy Causes Rage

"Backward countries get software communism before hardware, and this fills them with rage to see the discrepancy between their abundant software information environment and the lack of industrial services," he said.

Central to his theories—or "probes," as he calls them—is the insight that technology is profoundly reshaping modern man: Instead of "the alphabet and print technology," which fostered fragmentation, mechanization, specialization, detachment and privacy, we now have an "electric technology," which fosters unification, involvement, "all-at-once-ness" and a lack of goals.

In his view, the struggle to close the gap between software and hardware releases furious energies in tribal societies, in the Orient and in our children, who have been "tribalized" by absorbing TV (the new software) before books (the old hardware).

For this reason, he pointed out, we can expect growing violence from the young. "It's happening all over, from Columbia to Paris, and for the same reasons," he said. "We haven't seen anything yet. When the TV generation arrives, they're as likely as not to burn down every school. They won't hesitate to end the existence of cities either. They hate cities and machinery."

Intellectual Poverty of Schools

In his view, "the TV generation"—now 12 to 14 years old—is the first to have its sensibilities fully shaped by the new media and not by the older print-oriented, mechanical culture.

"All the young are in the same position as the Negro," he went on. "The discrepancy between the riches of the TV feast and the poverty of the school experience is creating great ferment, friction and psychic violence." But the new era—and "the new violence"—does not have an end in view, he said. "It is the process itself that yields the new image."

"When children go to school, they are filled with rage at the puny curriculum, the lack of information presented. The children in Watts were quite right in thinking, 'Why should we go to school to interrupt our education?' "

He views the Vietnam war as a struggle for identity on both sides. "The U. S. is like Don Quixote, fighting a war for a medieval image of itself. We have no goals in Vietnam, but an image we don't understand."

Part **2**

THE COURSE OF TECHNOLOGICAL CHANGE: PAST AND FUTURE

AMONG BOTH ITS critics and its advocates, the current and prospective state of technology is considered massive if not overwhelming. Both the range of technology and the speed with which it brings about change command attention. One encounters awestricken allegations that more than three-quarters of all scientists who have ever lived are now alive, or that the doubling-rate of knowledge has been reduced from millennia to around a decade. There are two points to be noted about this rapid acceleration. First, it did not just happen, or occur according to some natural law of growing acceleration. An expanding pool of information and principles does make possible an increasing number of new combinations, but neither increase is inevitable. The decision to invest in science and its applications mainly accounts for rapid change. Second, the confusion between science and technology, which occurs time and again in the essays reproduced here, rests upon an unspoken assumption that if a scientific discovery or correction of previous tenets has a potential application, the chances are fairly high that the application will occur. (Another aspect of acceleration is that the time gap between principle and practice also seems to be narrowing.) Yet science and technology are not an equation or simply a redundant set of syllables. I think the confusion is often made by scientists themselves, for in a society that values work and utility—and remains in that limited sense puritanical—it is just slightly discreditable to profess an interest in scientific or humanistic knowledge, or art, or play for their own sake.

We are warned, then, that when the authors of the following essays profess to speak for and about the past and future course of science, they are mainly concerned with applications, and that makes them useful here.

The first essay by Waldemar Kaempffert looks back a century (from 1951) and notes the failure of merely technical (read: physical or chemical) solutions to such social problems as poverty and war. Kaempffert's assertion that "In the past, science has taught man how to conquer his environment" would now be disputed, and his further assertion that "It [science] has still to tell him why he behaves as he does and teach him how to conquer himself" is, to put it gently, naive.

Another anniversary (Kaempffert's occasion was the centenary of the *New York Times*) was that celebrated by the Seagram Company in 1957, but the compiler of the various forecasts relating to 2057 does not tell whether they were written before or after a liberal sampling of the sponsor's principal products. The several scientists provide a taste of "futurology," which was later to become much more widely practiced. For those authors that worry at all about the predicted scientific and technical advance, the worry is consistently in the form of lagging social institutions.

For the sake of tragic rather than comic relief, I have next reproduced the essay by John Lukacs on George Orwell's anti-utopia, *1984*. The open or hidden purpose of anti-utopias is to predict a dismal future course in order to encourage prior preventive action. Incidentally, it is noteworthy that in Orwell's all too persuasive nightmare a form of psychological technology is used: the oppressed are finally made to love their oppressors.

The "futurologists" referred to above are the subject of William Honan's essay. The author sounds a cautionary note, not on the reliability of forecasting techniques as such, but to the effect that prediction of the future may be blinding or self-fulfilling.

Charles Frankel looks both backward and forward, and decries cheap "technical" means for understanding and controlling social institutions. The institutions *are* amenable to examination and predictive propositions, but ". . . it is a grave mistake to dismiss science as useless in solving moral and political problems."

The technical miracles that excite our other authors Arnold Toynbee takes for granted. What he finds noteworthy is the revolution—my term, not his—in both awareness of change and in the social conscience that attends to troublesome consequences.

The Past Century—
and the Next—in Science

by Waldemar Kaempffert

A CENTURY has elapsed since *The New York Times* first appeared in 1851—a century of revolutionary changes in scientific outlook and in the environment beyond the imagination of a Jules Verne. If a well-read business man, lawyer or banker who was in his prime in 1851 could be resuscitated and dropped in our midst he would be amazed, puzzled and bewildered by what he saw, heard and experienced. He would wonder what a reporter meant who, in summarizing the proceedings of a medical meeting, referred to "vitamins," "hormones," "antibiotics." There were no such words in 1851 because the compounds for which they stood had not been discovered. Pasteur and Koch had not yet done their work, so that very little was known about the mechanism of bacterial infections and still less about that of diseases caused by viruses.

The physician of 1851 was no ignoramus, but he had only a smattering of what we would call "scientific medicine." There was no attempt to measure blood pressure in the routine practice of medicine, no electrocardiography of the heart's action, no way of examining the chest and the stomach with X-rays because Roentgen, their discoverer, was only 6 years old. What progress in medicine has meant is shown by the life tables of insurance companies. A white boy baby born about 1850 had what statisticians call a "mean expectancy of life" of 41.8 years, a white girl baby of 44.9. The corresponding figures for 1948, the latest year for which they are available, are 65.49 and 71.04 years.

The lights of Main Street would dazzle that citizen of the mid-nineteenth century; the absence of horses in the streets would call for comment; the automobiles that sweep silently past would astound and frighten him; the airplanes overhead would arouse wonder; a man in a telephone booth holding one end of a black instrument to his ear and talking into the other end would mystify him; motion pictures, radio broadcasting and television would entertain him but puzzle him, too, because he would not understand the underlying principle of their operation. Skyscrapers would seem fantastic and not the economic necessities that they are.

The world of 1851, out of which that resuscitated citizen came, was much more self-satisfied than ours, so far as scientific beliefs are concerned. Newton's laws were considered inviolate. Light was conveyed to us from the sun or from a candle as a series of ripples in the ether, a medium that was supposed to pervade everything and that was more tenuous than any gas, yet more rigid than steel. Atoms were the smallest indivisible units of which matter was composed. The only disturber of mental peace was Charles Darwin, whose "Origin of Species" (1859) and "Descent of Man" (1873) at first shocked the Western world but ultimately made it necessary to regard all living creatures as the warp and woof of a single fabric.

Nearly all the fundamental scientific conceptions of a century ago have been either shattered or modified, Darwin's included. Natural selection, a scythe, did not explain how new species originated, but de Vries and others who developed the mutation theory in the early part of this century did. The anthropologist of our day believes not that man is a descendant of an Old World anthropoid ape, as Darwin concluded, but that apes and men are branches of the same limb of the family tree.

As for our conception of the world around us, Albert Einstein, a young examiner in the Patent Office of Switzerland, evolved a theory of relativity in 1905 and 1915 which showed that Newton's laws did not conform with reality. He swept the ether away. Max Planck convinced physicists that light and radiation in general came not in ethereal waves but in packets called "quanta." Henri Becquerel in 1895 accidentally discovered that uranium was radioactive.

Another accident led Wilhelm Konrad Roentgen in the same year to discover that from one electrode of an evacuated tube through which an electric current passed came invisible rays that would penetrate flesh and reveal the bones of the body. In 1897 J. J. Thomson of Cambridge found that the current in such a tube was composed of electrons, each of which was only 1,840th as big as the hydrogen atom, so that the mid-century notion of an atom as the smallest possible chemical unit collapsed. In 1898 the Curies discovered in radium an element even more powerfully radioactive than uranium. Like uranium this radium shot out bits of itself that were smaller than atoms.

Because of these discoveries a new conception of the atom was necessary. It came from Ernest Rutherford, Niels Bohr and others. Textbooks on physics had to be rewritten and physicists lost their old, jaunty cocksureness. In 1851 everything was interpreted in terms of mechanism and laws of nature; in 1951 chance rules in the universe, and the laws of nature are recognized for what they are—mere statements of statistical averages.

There are now doubt and uncertainty in physics. Scientists of the Eddington-Jeans school studied equations that were supposed to reveal the secrets of the atom, and hence of reality, only to find that reality had vanished and that trees, houses and stars were not what they seemed to be but only indications of "pointer readings" of a deeper something that was real—something that science could never reach. Is it just a coincidence that the same uncertainty prevails in art, in economics, in international relations?

The changes in outlook brought about by research in fundamental science are of more practical importance than is at first apparent. Mendel's laws of heredity, when they were rediscovered in 1900 and later amplified by a score of geneticists, made it possible to breed animals and plants almost to commercial specifications.

And out of all the studies of radioactivity and the formulation of new theories of atomic structure came a way of releasing energy from the nucleus of an atom—the supreme scientific achievement of man. Stars and atoms, relativity and nuclear physics, are linked together. Out of this linkage came electronics, a branch of engineering that has grown up within the memory of middle-aged men.

Engineers of the late nineteenth century designed electrical machinery without ever knowing what electricity was. After Thomson, Becquerel and the Curies they knew. A current in a wire was a continuous flow of electrons, a lightning stroke a gush of them. With this new knowledge came the electronic or vacuum tube which has ten thousand uses. One of the uses is the conversion of light into electrical current or of an electric current into light, a conversion that made television possible.

Now engineers are applying electronics in machines that solve in a few minutes problems that would ordinarily keep a mathematician busy for weeks and months. The creators of the electronic computers and other variations of "thinking" machines are certain that in the factory of the future more machines with the skill of intelligent craftsmen will be seen. Holes punched in ribbons of paper will instruct and operate the machines. Anything will be produced automatically from an electric toaster to an automobile.

As we look back at a century of technologic change, from the standpoint of recent electronic triumphs, three periods of invention can be distinguished.

In 1851 the world was still living in the first period, the period inaugurated by the introduction of the steam engine—the period of steam-driven machinery on land and sea. The telegraph was the only electrical invention in wide use until the Eighties, and the most potent source of electricity was a battery of 20,000 cells owned by the Western Union Company.

The second period begins with Thomas A. Edison—the period that the romanticists of the Eighties and Nineties called "the electrical age." In the history of invention no figure towers above his. He was just four years old in 1851. It was he who built the first steam-driven central power station; he who invented an electric lamp that was equally useful in the home, the factory or the street; he who devised electric motors that could drive factory machinery or trains. Fifteen hundred inventions are associated with his name.

Though he lived well into the twentieth century Edison belonged to the nineteenth. There will always be lone, garret inventors, but he was probably the last colossus of invention. In his old age his place was already taken by the industrial laboratory, in which trained physicists, chemists and engineers worked in groups. No single genius could solve the problems presented by a modern telephone system, by an atomic bomb or by the complicated electronic devices of today.

With the development of the electron theory of matter we entered the third, or electronic, period of invention. Not only skill but something easily mistaken for intelligence has passed to the electronic machine.

With the evolution of the three periods of invention there has also been an evolution in our sense of social responsibility. A century ago a few historians and philosophers saw that a revolutionary process or a machine had social effects. The business man did not worry much over such matters as the displacement of labor that would follow the introduction of an invention or the disfigurement of the countryside by heaps of slag, or the menace to health caused by chimneys that belched smoke all day long and sometimes far into the night. As for the deeper effects of progress in science and technology—the change in folkways and in modes of life—little attention was paid to them until the beginning of the present century.

There is not an industry, not a vocation, not a mode of life but has felt the impact of science and technology. Old skills have been abandoned and new ones, highly paid, have been acquired. The enormous gap that separated the very rich from the very poor in 1851 has been narrowed by science as much as by heavy taxation of the rich. This broader distribution of wealth has broken down social distinctions. Millionaires and machinists are indistinguishable on Broadway. Where are the "400" of Ward McAllister? Their places have been taken by "café society," a term that speaks for itself.

Family ties are looser than they were in 1851, and for that the motion picture and the automobile are partly responsible; also the migration of women from the home first to the factory, later to the office when the typewriter and other office appliances were introduced. A century ago many Americans were their own masters in the sense that they were craftsmen who owned their own shops or business men who started with only a few hundred dollars to make shirts or print books. Today we are more a nation of employes rather than of self-employers, because it costs millions to build and equip a plant in which steel is produced, automobiles are assembled or textiles are woven.

It was assumed in 1851, as it was assumed a century earlier, that institutions and folkways would remain as they were regardless of the advance of scientific technique. But Eli Whitney's gin was more than a device for picking seeds out of cotton; it revived the moribund institution of slavery and prompted the Republican party to formulate a high-tariff policy, which would force Southern planters to send their cotton to New England textile mills instead of to England. The automobile, similarly, proved to be more than a substitute for the horse and carriage, radio more than a means of dispensing with wires in electrical communication, the motion picture more than an ingenious way of showing nature in action.

Why do scientific discoveries and inventions have this social effect? Because we cannot make use of a new discovery or invention unless we adapt ourselves to it. The more revolutionary the discovery or invention the more urgent is the need of adaptation. The introduction of the fountain pen did not demand any change in our writing habits. On the other hand, the introduction of the automobile made it necessary to spread a network of fine roads over the entire country; and with the roads came tourist camps, "motels," wayside restaurants and less pretentious hamburger stands.

Twenty miles was a long journey in 1851. Today thousands live twenty miles from the nearest town and think nothing of traveling in their own cars across the continent. The motion picture throttled the theatre and created Hollywood and with it an industry that is capitalized at billions and that entertains millions every day. Push-button electric conveniences have solved the servant problem, just as canned and frozen foods have solved the cooking problem.

Back in 1851 and long thereafter a maid worked from 6 or 7 in the morning until 10 or 11 at night. For the last ten years she has usually been hired by the hour at a wage that many a country school teacher would be glad to earn. The permanently employed housemaid may be a passing phenomenon. Even today women whose husbands earn $25,000 a year and more do their own housework—but with the aid of electrically driven machinery. Changed economic conditions explain what has happened in the home and give inventors their chance, but the change in economic condition and hence in industry is itself the result of the change that has occurred in science and technology.

If the citizen of 1851 was still much of an individualist, science and technology have transformed his descendant of 1951 into a consumer of mass-produced goods. Mass production is the distinguishing characteristic of our time. Without standardization, mass production as we know it is impossible. Hence the uniformity that prevails.

One town looks very much like another through the windows of a railway train. From New York to San Francisco grocers stock identical standardized packages of cereals, identical standardized cans of vegetables, identical standardized packages of frozen food. The electric lights of one Main Street are exactly like those of ten thousand other Main Streets. Very little latitude is allowed in the selection of apparel, especially men's apparel. It is harder than ever to be "different." Choice in recreation, choice in what we eat and wear, choice in everything is restricted.

Even life in the rural districts has been standardized. The farmer of 1851 was a hayseed, a yokel who seemed out of place in a big city. His counterpart of 1951 rides to town in his own car, goes to a movie, shops in a department store. So far as appearances go he hardly differs from the salesman who waits on him. He may grow wheat but, like anyone who lives in a city, he buys flour in a sack, breakfast foods in cardboard containers, vegetables in cans. In 1851 there were still flour mills to which a farmer could take his grain to be ground. Gone is the isolation of 1851. A radio set tells a farmer what is going on in the world. The telephone enables him and his wife to gossip with friends who live ten miles away, friends who are electrical neighbors.

Standardization and the collective utilization of energy have had the effect of knitting the population of a huge community together. A metropolis is a colossal organism, which is composed of such elements as skyscraper office buildings, apartment houses and hotels, subways, telephones, water and gas mains. Millions in homes, factories and offices are enmeshed in the hidden wires of a single central electric powerhouse. Let an accident occur, like one of about twenty years ago that crippled a substation on New York's East Side, and lights go out, elevators stop running, ice cubes are no longer made in refrigerators, vacuum cleaners are impotent, candles have to be burned, homes and offices are cut off from one another, each fending for itself.

Without the telephone the skyscraper would be impossible; with the telephone metropolitan cities like London, Paris, New York and Chicago develop into regions. Some remnants of sectionalism remain in the United States, but when ten million persons scattered between San Francisco and New York or St. Paul and New Orleans listen to the same radio program, or when one friend in Boston telephones another in New Orleans to wish him a Merry Christmas, the effect cannot but be one of unification.

This mechanization and electrification of life, this onward sweep of standardization and the collective utilization of energy demands planning, organization and competent direction. Who are the planners, organizers and directors? Scientists and engineers in industrial laboratories, corporation executives, bankers and Government technologists. A new class has arisen, a class of expert planners, designers and managers, that holds its place through sheer ability, unlike the old hereditary military caste of nobles, and that rules industrial empires unimagined in 1851— empires of oil, chemicals, synthetic fibers, electric communication.

These experts at the top, on whom we are utterly dependent, are the molders of our scientific and technologic culture. There are not many of them—perhaps not more than 50,000. If they were to disappear we should be helpless until we had trained their successors. They are the instruments of Progress—spelled with a capital—about which so much was written in the Eighties and Nineties, especially by such evangelists of science as H. G. Wells. Give us more machines to do the backbreaking, grimy work of the world, ran the formula, and there will be an end of misery and poverty; give us more international means of mass communication, like radio and motion pictures, and alien peoples will understand one another, with the result that there will be no more wars; give us more science and more international scientific congresses, and nations will learn to sink their differences in the common cause of enlightening one another.

None of these predictions has been fulfilled. The world has never been so restless, so uncertain of its future, so terrified at what may happen if another world war is fought with all the aid that science can lend. Every advance in science and technology has both improved man's material lot and heightened his military power. According to Pitirim A. Sorokin of Harvard 957 important wars had been fought between 500 B.C. and 1925 A.D., and we are now living in what he calls "the bloodiest crisis of the bloodiest century."

It is an old complaint that man's increasing technological ingenuity has not been accompanied by a corresponding sense of moral obligation, meaning that we have not yet succeeded in curbing abuses of scientific discoveries and inventions either in peace or in war.

About forty years ago, when we first began to dream of releasing atomic energy, Sir Oliver Lodge doubted whether man was spiritually and morally fit to make rational use of such a technical triumph. The same note is struck today. With the introduction of electronic machines that will do much of man's thinking and production for him, Dr. Norbert Wiener, who has had much to do with their recent development, is not convinced that we will do any better with them than we did when we permitted slums to grow around steam-driven factories.

Science has so far been applied chiefly in fathoming the secrets of matter and motion, technology chiefly in winning wars and making profits, changing the environment and improving man's material condition. If the proper study of man is mankind, science is still in a primitive state. It knows less about man, the most interesting and important object in the universe, than it does about the atom.

Science needs another Renaissance. It may dawn in another century. If it does, science will devote at least as much research to man's abilities and shortcomings as it has devoted to the synthesis of drugs and plastics. In the past science has taught man how to conquer his environment. It has still to tell him why he behaves as he does and teach him how to conquer himself.

Science Looks at Life in 2057 A.D.

WHAT IS *in store for man, in view of the enormous forward strides science is making? Not long ago, in New York, a panel of outstanding scientists offered their answers, projecting themselves a century into the future in a symposium on "The Next Hundred Years." The occasion was the centennial celebration of Joseph E. Seagram & Sons, Inc. Below are excerpts from some of the scientists' forecasts.*

"Predictable Progeny"

Provided that the world does not fall prey to one of the four dangers of our times—war, dictatorship of any kind, overpopulation or fanaticism—the coming one hundred years will revolutionize advances in sciences and their application. For example, it will become possible to bring the reproductive cycle under regulation, to prescribe the sex of the child and to produce at will twins either identical or fraternal, or still more multiple births.

The prevention of overpopulation can occur only by the widespread acceptance of the philosophy that the number of offspring to be produced should be voluntarily restricted for the good of the offspring themselves. With this more ethical attitude concerning reproduction, it will also be regarded as a social obligation to bring into the world human beings as favorably equipped by nature as possible, rather than those who simply mirror their parents' peculiarities and weaknesses.

Foster pregnancy, already possible, will be readily achieved and widely welcomed, in addition to natural pregnancy. This will provide the opportunity of bearing a child from the union, brought about under the microscope, of productive cells derived from persons who exemplified the considered ideals of the foster parents.

The reproductive cells will preferably be derived from persons long deceased, to permit a better perspective on their work, relatively free from personal pressures and prejudices. For this purpose banks of frozen reproductive cells, such as we already have today, will be maintained.

Even more predictability concerning the nature of the progeny will be attainable, if desired, by a kind of parthenogenesis. Where offspring now ordinarily have their hereditary material picked in a random way from two different parents, in this case the offspring would obtain his hereditary equipment entirely from one individual with whom he is as genetically identical as if he were an identical twin. This will be accomplished, as is now done, by extracting the nucleus from a human egg and inserting in its place an entire nucleus obtained from a cell of some pre-existing person, chosen on the evidence of the life he or she had led, and his or her drive potentialities.—DR. HERMAN J. MULLER, *geneticist, Nobel Prize winner, Professor of Zoology, University of Indiana.*

"Cosmic Wonders"

I believe the intercontinental ballistic missile is actually merely a humble beginning of much greater things to come.

Let us suppose it is now 2057 A.D.—one hundred years from now. The wonders of the cosmic age have unfolded before the eyes of mankind. Several expeditions have already gone to Mars and Venus and exploratory voyages will have been extended as far as Jupiter and Saturn and their natural satellites.

Voyages to the moon have become commonplace. The surface of the moon has been subdivided into spheres of interest by the scientists of the major powers, and a lot of prospecting, surveying, tunneling and even some actual mining of precious ores and minerals are going on.

At some particularly scenic spot in the moon, lavish excursion hotels have been established. These are operated by several national space lines for the purpose of attracting more passenger traffic in addition to their main business of hauling commercial cargo. All these places are pressurized and air-conditioned, featuring large picture windows and astral domes to do justice to the magnificent scenery. They run the whole gamut from honeymoon hotels to wide open gambling joints.

Transportation costs to the moon and to the planets have been immensely reduced as a result of the replacement of the early chemically powered rocket ships by ships powered with controlled thermonuclear energy. Definite plans are under way for a regular transport system between the earth and the near planets.

The earth, in 2057 A.D., is surrounded by a whole family of artificial satellites, all of them accepted as members in good standing of our solar system. They are in a great variety of sizes, brightness, purposes, nationalities, orbital altitude and orbital inclinations. Some of the satellites, the best moneymakers, have taken over the mailman's job. They receive messages radioed up to them while over one city, country or continent, and play them back while over another.

A few such communication satellites will handle the entire volume of private and official communications between all points of the earth which are more than five hundred miles apart. No message will require more than one hour from sender to recipient. Other satellites at various altitudes will serve as television stations for nation-wide and global television. In addition, there will be large manned space stations serving as space terminals for voyages to the moon and the planets.

When, in 1957, one hundred years ago, the world was rudely awakened to the dawn of the cosmic age by the Russians' sputniks, America momentarily was caught napping, and swung into energetic action. It did not take long for most people in this country to realize that there was far more behind the sputniks than the sensational "beep-beep" and the space-ferrying dog. Indeed, was it not all too obvious that the Soviet Union of 1957 looked upon outer space very much as, in the seventeenth and eighteenth centuries, Great Britain looked upon the seas?

To build a world-wide empire, Britannia then had to rule the waves. To control the great globe itself, the Soviets had to control the space around it. But, fortunately, they didn't make the grade after America, the napping industrial giant, got into the act. Thus, in 2057, the sun still rules over the cosmic age.—DR. WERNHER VON BRAUN, *rocket expert, Director of the Army Ballistic Missile Agency, Huntsville, Ala.*

"Golden Age"

I cannot but believe that the next one hundred years will prove to be more critical than any mankind as a whole has yet been called upon to face. If we survive the next century, and if we are successful in preserving our industrial civilization without becoming robots in the process, then I believe that truly wondrous vistas of our world and of our universe will present themselves in endless sequence.

As our supplies of oil dwindle in this century and coal in the next, we will shift to nuclear power. The world supply of atomic energy is almost inexhaustible. There are large reserves of uranium to start off with and, when these have been consumed, we can, if necessary, obtain uranium from ordinary rock. After that, we have the vast potentialities of thermonuclear power.

A world-wide golden age is truly within our reach. The unknown factors in the equation are not the potentialities of science and technology. The major unknown, I believe, is whether man can devise the moral, the social, and the political means of living with man—before it is too late.—DR. HARRISON BROWN, *Professor of Geochemistry, California Institute of Technology.*

"Vegetable Steaks"

Before the year 2057 we will probably understand in the most meticulous detail all of the molecular and atomic events that cause living things to live.

It appears to me quite probable, however, that people at this time will still eat food. By this I mean that I consider it unlikely that human beings will take on their supply of energy directly as electrical current or as nuclear power. It is widely held that we will one day replace food with a pill. Perhaps. But if so, I think it will be a big pill. It will be approximately the size of a present-day meal rolled up into a ball. This food will continue to come in the main from green plants, as it does today, from plants that are grown for the purpose by agriculture.

Since there will be several times as many people in the world in a hundred years as there are today, it will be necessary to have, all over the earth's surface, an agriculture that is more intensive than that we now know. Most of the earth's surface will be tilled as intensively as is now done in Japan or Denmark, and to do that we will have to extend our cultivated areas. We will irrigate the deserts with water purified from ocean water. Loss of crops to pests will be long abolished and be just a dim memory.

Farming, of course, will be very highly mechanized and very few people will be directly involved in it. It will be possible, in fact, to program the entire farming operation and leave the farm to run itself from a master computer panel.

Human beings one hundred years from now will be largely

vegetarian. As the mass of human flesh increases on the earth's surface, the mass of animal flesh will inexorably decrease. But our vegetarian diet will be a wholly satisfactory one, nutritious, attractive and wholesome. The craft of food technology will have reached a high level. We will, for example, eat steaks made from extracted vegetable protein, flavored with tasty synthetics and made chewy by addition of a suitable plastic matrix.—DR. JAMES BONNER, *Professor of Biology, California Institute of Technology.*

"Outdoing Nature"

I would like to pose two questions: What makes the grass green? How do oysters get their copper out of the sea water in which they live?

I am very serious about these questions. They involve the complex details of the most minute chemical mechanisms within the cells of living matter. They lie in the general domain of the biochemist.

The first deals with photosynthesis, which is that process by which plants absorb the energy of sunlight and use it to bring about the energy-absorbing chemical reactions between carbon dioxide and water and a few other materials to form the substances of which plants are made. This process is basic to all living matter, including ourselves.

The chemist happens to know a great deal now about photosynthesis, but much knowledge remains to be acquired. When this subject is mastered, the chemist, aided by the physicist and the chemical engineer, will be able to develop processes of artificial photosynthesis; this will open up the possibility of synthesizing all of our liquid fuel, such as gasoline, not just by imitating, but by outdoing nature.

Taking up the second question, we know that the oyster gets its copper from the sea water in which he lives, and we know that he has concentrations of copper in his blood-like fluid, a few thousandfold greater than in the water which surrounds him.

How does he do this? As far as I know, no one knows the answer. The process is understood in a general sort of a way, but the details still remain a mystery. When that process is understood in detail, then we will be able to improve on nature, and I think it may very well be that we will be able to develop processes for tapping the almost infinite mineral resources of the ocean.

The questions that I posed are only two of a very great many which fall in the domain of the biochemist. What is the biochemistry involved in the life process itself? What is living matter? How would such knowledge, if we had it, affect the specific cure for cancer, for instance? Almost certainly, there are some biochemical abnormalities involved in mental diseases. What are they? Can those mechanisms be altered? There are many more such unanswered questions.

The coming century, in the hands of the biochemist, may very well be the dominant period of progress of the world—*provided* the proponents of the other scientific disciplines, the politicians, the public at large, seriously absorb and use the discoveries which, I am sure, the biochemist, with his colleagues, is going to make.—DR. CLIFFORD C. FURNAS, *chemist, Chancellor of the University of Buffalo.*

"A Delightful World"

If man survives, he can look forward to learning more about himself in the next one hundred years than he has in the preceding one million. He could discover the causes and cures of sicknesses and pain, of hate and destruction. He could come to realize his true biological potentials. He can learn to circumvent many of his limitations. He can learn how to change himself. Conceivably, in another hundred years, the life sciences and social sciences could put man in control of his own destiny.

The chemistry of metabolism and body processes will provide a detailed knowledge of the biochemistry of the central nervous system. It might then be possible to change the environment of the nervous system and thus eliminate disease and malfunction and produce an increased biological efficiency of nerve cells and of the cell network. We would then be able to change emotions, man's desires and thoughts, by biochemical means, as we are now doing, in a rather gross way, with tranquilizers.

The genesis of human motives, values, feelings and emotions, and the way in which our child-raising procedure influences our development, should be worked out in detail. Then we would know how parents could provide an ideal environment for their children to grow into emotionally secure, self-confident, happy adults. Mental disease, emotional illness, neurosis, maladjustment, psychological insecurity would then be eliminated.

Certainly, there will be new problems of mental and emotional adjustment that will emerge as our society becomes more complex and more demanding. But these malfunctions should be on a more and more superficial level as we continue to work out the basic principles of human behavior.

The intellectual output of the brain should be greatly increased. The important principles concerning the thinking processes, as they relate to creative imagination, will be worked out, and the procedure will be so systematized that man should be able to generate creative ideas at will, simultaneously taking into account all possible combinations of known variables.

Educational practices should be radically different. They should depend much less on verbal communication and more on the other senses. Electronic memory banks, complex computers, perhaps even coded electrical information transmitted directly into the nervous system—all of these could accelerate formal education.

I think we might expect parallel advances in uncovering the principles of dynamics; that is, the network of psychological and emotional forces that influence the behavior of people in group and social situations. These principles would tell us how to form groups, how to develop group goals, how to select group leaders, how to reach effective group decisions.

The process by which an aggregate of people becomes a closely knit unit, an integrated team, will be understood. This will enable us to make very rapid social changes, to eliminate the lag in culture, and to develop desirable social organizations in relatively very short spans of time.

And so, man will continue to increase his intellectual resources to meet the demands of the future. If he learns to control himself before he destroys himself, he will find the world a delightful place in which to live.—DR. JOHN WEIR, *Associate Professor of Psychology, California Institute of Technology.*

It's Halfway to 1984

by John Lukacs

WE ARE NOW halfway to 1984. George Orwell, the author of "1984," finished his book in 1948. That was 18 years ago, and it is not more than another 18 years before that ominous date rolls around.

It is *ominous,* in every sense of that antique adjective. There is reason to believe that 18 years from now thousands of people will experience a feeling of uneasiness, perhaps a light little shudder of trepidation, as they first encounter that new year's numerals in print. In the English-speaking world, at least, "1984" has become a household term, suggesting some kind of inhuman totalitarian nightmare. And since millions who have not read the book now recognize the term, it is reasonable to assume that both the theme and the title of the book have corresponded to an emerging consciousness among many people in the otherwise progressive-minded English-speaking democracies, to the effect that things are *not* getting better all the time—no, not at all.

The plot of "1984" is well-known but it may be useful to sum it up briefly. By 1984 most of the world has been divided by three superstates—Oceania, Eurasia and Eastasia. They are perpetually at war with one another, but no one of them is completely able to subdue the others. This state of war enables the rulers of these states (the ruler of Oceania being Big Brother) to keep their peoples both ignorant and submissive. This is achieved by totalitarian and technical methods, by the absoluteness of one-party rule and by a kind of censorship that controls not only the behavior but even the thinking process of individuals. The hero of "1984," Winston Smith, born in 1945 (both the date and the first name are significant), is a simple party member and a functionary of the Ministry of Truth in London, which is the chief city of Airstrip One, for that is what Britain became after she had been absorbed by the United States to form Oceania. (Continental Europe, having been absorbed by the Soviet Union, had become Eurasia.)

Winston is a weak and forlorn intellectual who, however, is sickened not only by the dreary living conditions in 1984 but by the prevalence of official lying and the almost complete absence of personal privacy. One day he stumbles into a love affair, which in itself is a dangerous thing since the party punishes illicit relationships severely. Winston experiences happiness and a sense of personal fulfillment, especially as Julia shares his hatred of the existing system.

There is a high official in the Ministry of Truth, O'Brien, whom Winston instinctively trusts. He and Julia confide in O'Brien. They are deceived. All along, O'Brien has set a trap for them: they are arrested in their secret little room. They are tortured. Winston, despite his strong residue of convictions, not only confesses to everything imaginable, but in the end, faced by an especially horrible torture, he even betrays Julia. He is finally released; he is a completely broken man; he has even come to believe in the almightiness and goodness of Big Brother.

But it is not this plot, it is rather Orwell's description of everyday life in 1984 that is the principal matter of the novel and, one may suppose, the principal matter of interest to its readers. Life in 1984 is a mixture of horror and dreariness. What is horrible is not so much war as the shriveling of personal freedoms and privacy with the planners of the superstate controlling vast portions of once-independent lives. What is dreary is that within these totalitarian conditions the living standards of masses of people in what were once civilized and prosperous countries are reduced: Food and drink are little better than standardized slop; mass entertainments are primitive and vulgar; personal property has virtually disappeared.

One of the profound differences between "1984" and Aldous Huxley's "Brave New World" (published in 1932, the latter still had many of the marks of the light-headed twenties; its philosophy compared with that of "1984" is a rather irresponsible *jeu d'esprit*) lies in Orwell's view of the past rather than of the future. Looking back from 1984, conditions in the early, capitalistic portion of the 20th century seem romantic and almost idyllic to Winston Smith, so much so that on a solemn occasion he offers a toast "to the past." Unemployment, revolutions, Fascism and, to some extent, even Nazism and Communism are lesser evils than what is going on in Oceania in 1984, since by that time the rulers of the state have perfected brainwashing and thought-control to the point that the memories of entire generations, and hence their opinions about the past, have been eliminated.

This, of course, does not happen overnight: It is a brutal but gradual development. In "1984," Orwell set the decisive turning point in the middle sixties, "the period of the great purges in which the original leaders of the Revolution were wiped out once and for all. By 1970 none of them was left, except Big Brother himself."

Let us keep in mind that "1984" is the work of a novelist and not of a prophet; Orwell ought not be criticized simply because some of his visions have not been borne out. On the other hand, Orwell was concerned in the late forties with certain tendencies of evil portent; and "1984" was a publishing success because around 1950, for great numbers of people, the picture of a society such as he described was not merely fantastic but to some extent plausible.

It is still plausible today, but not quite in the way in which Orwell envisaged the future 18 years ago. Halfway to 1984 we can say, fortunately, that most of Orwell's visions have proved wrong. It is true that the United States, the Soviet Union and China correspond to some extent to the superpowers Oceania,

Eurasia and Eastasia. But the United States has not annexed Britain, the Soviet Union has fallen far short of conquering all of Europe, and even China does not extend much beyond her traditional boundaries.

What is more important, the superpowers are not at war with one another. It is true that during the so-called cold war between the United States and the Soviet Union many of the practices of traditional and civilized diplomacy were abandoned; but the cold war has given place to something like a cold peace between these two superpowers. Even the dreadful and ominous war in Asia is marked by the reluctance of the United States and China directly to attack each other.

Orwell proved correct in saying that "war . . . is no longer the desperate, annihilating struggle that it was in the early decades of the 20th century. It is a warfare of limited aims between combatants who are unable to destroy one another. . . ." Yet Orwell was interested principally not in international but in internal developments. For example, in "1984" the peoples of Oceania are isolated; travel is forbidden except for a small minority of the élite; and the press is controlled to the extent that no meaningful information from the outside world is available to the public.

But now, halfway to 1984, the opposite has been happening. It is not warfare but torrents of automobiles and mass tourism that threaten to destroy entire landscapes and cityscapes; great amounts of information are available to us about an undigestible variety of matters; and at times it seems that the cultural traditions of great Western nations are endangered less by the persistence of isolationism than by a phony internationalism drummed up by a kind of pervasive publicity that drowns out the once truer music of the arts.

Also, in the world of "1984" most people are ill-fed, badly clothed, run-down. But this, too, has not happened. Now, halfway to 1984, almost everywhere in the world, living standards have risen, and the danger is not, as Orwell envisaged it, that entire generations of once-prosperous countries will no longer know such things as wine, oranges, lemons and chocolate; it is, rather, that our traditional tastes and table habits may be washed away by a flood of frozen and synthetic foods of every possible kind, available to us every hour of the day.

The reasons why Orwell's visions of 1984 have been wrong seem to be bound up with the time and the circumstances of the book's conception. About the circumstances Orwell himself was supposed to have said that "1984" "wouldn't have been so gloomy if I had not been so ill." He wrote most of the book in self-imposed isolation on a rain-shrouded Scottish island, finishing it in an English country hospital in late 1948. Shortly thereafter, he was moved to a hospital in London, where in January, 1950, he died. As for the time of writing, in the late nineteen-forties Orwell's imagination succumbed, at least in part, to the temptation of conceiving the future as an increasingly acute continuation of what seems to be going on at the present. (In one of his earlier essays, Orwell had criticized the American writer James Burnham for this very fault.) Around 1949, when most intellectuals had come around to recognizing that Stalin's tyranny was hardly better than Hitler's, many of them concluded that it is in the nature of totalitarianism to become more and more tyrannical as time goes on. Indeed, some of them established their reputations by the ponderous books they produced on this theme. (Hannah Arendt's "The Origins of Totalitarianism" is an example.) Yet only a few years later, events in Eastern Europe and in Russia showed that history is unpredictable and that the projections of intellectuals are often oversimplified. But this Orwell did not live to see.

He foresaw the horrible features of 1984 as the consequences of totalitarianism, of political tyranny, of the despotism of a dictator. But halfway to 1984 we can see, for example, that the era of totalitarian dictatorship is sliding away, into the past. Even the Soviet Union seems to be moving in the direction of what one may call "post-totalitarian"; all over Eastern Europe (though not yet in Asia) we can perceive regimes that, though dictatorial, are no longer totalitarian. The danger for us is, rather, the obverse: the possibility of totalitarian democracy.

Totalitarian democracy? The words seem paradoxical; our eyes and ears are unaccustomed to the sight and the sound of them in combination. Yet I believe that we ought to accustom our imaginations to the possibility of a democratic society in which universal popular suffrage exists while freedom of speech, press and assembly are hardly more than theoretical possibilities for the individual, whose life and ideas, whose rights to privacy, to family autonomy and to durable possessions are regimented by government and rigidly molded by mass production and by mass communications.

Let me, at this point, fall back on a personal illustration. For a long time the term "1984" evoked, to me, the image of a police state of the Eastern European type. But when I think of 1984 now, the image that swims into my mind is that of a gigantic shopping center and industrial complex—something like the one which has been erected a few miles from where I live in eastern Pennsylvania.

The undulating rural landscape around Valley Forge, with its bright dots of houses and its crossroads, has been transformed. There is now the eerie vastness of the General Electric Space Center whose square edifices spread across hundreds of acres. Beyond it stand other flat windowless blocks of buildings—the King of Prussia shopping center, around the trembling edges of which bulldozers roar from morning to night, boring their brutal tracks into the clayey soil which they must churn to mud before it can be covered by concrete. The predominant material is concrete, horizontal and vertical concrete. Twice a day, thousands of people pour into and out of this compound, in a tremendous metallic flow. But no one lives there. At night and on Sundays, these hundreds of acres resemble a deserted airport, with a few automobiles clustering here and there, or slowly cruising on one of the airstrips, occasionally peered at by uniformed guards. Why fly to the moon? Stand on a cold January night in the middle of a parking lot in a large shopping center in the American North. It is a man-made moonscape. This is how the moon will look after our Herculean efforts, after we reach it, colonize it, pour concrete over it.

This is how 1984 looks to me, in the middle sixties, but I know and feel that this view is neither solitary nor unusual.

There are millions of Americans who, passing a similar space-age complex of buildings, will say "1984," covering up their resignation with a thin coat of defensive humor. What strikes us is not just the ugliness of the buildings but something else, something that is not so much the reaction of middle-aged earth-men against brave new worlds as it is the expression of a feeling which is, alas, close to the Orwellian nightmare vision: a sense of impersonality together with a sense of powerlessness.

The impersonality is there, in the hugeness of the organization and in the anonymous myriads of the interchangeable human beings who make up most of the personnel. The powerlessness is the feeling which I share with so many of my neighbors—that we cannot stop what in America is called the March of Progress, the cement trucks coming toward us any day from across the hill; the knowledge that our voices, our votes, our appeals, our petitions amount to near-nothing at a time when people have become accustomed to accepting the decisions of planners, experts and faraway powerful agencies. It is a sickening inward feeling that the essence of self-government is becoming more and more meaningless at the very time when the outward and legal forms of democracy are still kept up.

Let us not fool ourselves: Now, halfway to 1984, with all of the recent advances of civil rights, with all of the recent juridical extensions of constitutional freedoms, we *are* facing the erosion of privacy, of property and—yes—even of liberty. This has nothing to do with the Communist Conspiracy or with Ambitious Government Bureaucrats—that is where our New Conservatives go wrong. It has nothing to do with Creeping Socialism. It has very much to do with Booming Technology. The dangers which our modern societies in the West, and particularly the United States, face now, halfway to 1984, are often new kinds of dangers, growing out of newly developing conditions. What ought to concern us is the rootlessness of a modern, technological, impersonal society, with interchangeable jobs and interchangeable people, on all levels of education.

We ought to dwell less on the possibility of unemployment arising out of automation, in a society which, after all, feels

obligated to produce full employment; rather, we ought to consider the growing purposelessness of occupations in a society
where by now more people are employed in administration than
in production. And in such a society we ought to prattle less
about the need for more "creative leisure" when the problem is
that work becomes less and less creative. We ought to worry
not about the insufficient availability of products but about the
increasing impermanence of possessions. We ought to think
deeply not so much about the growth of the public sectors of
the public economy at the expense of private enterprise (which,
at any rate, is no longer very "private"), but rather, about the
cancerous growth of the public sectors of our existence at the
expense of the private autonomy of our personal lives.

We ought to concern ourselves less with the depreciation of
money and more with the depreciation of language; with the
breakdown of interior, even more than with the state of exterior,
communications—or, in other words, with the increasing practices of Orwell's Doubletalk and Doublethink, and with their
growing promotion not so much by political tyrannies as by all
kinds of techniques, in the name of Progress.

I cannot—and, perhaps, I need not—explain or illustrate these
concerns in greater detail. They are, in any event, 1966 concerns about the future, not 1948 ones. Still, while many of the
phantoms that haunted Orwell's readers 18 years ago have not
materialized, the public currency of the term 1984 has lost none
of its poignancy. The tone of our literature, indeed of our entire
cultural atmosphere, is far more pessimistic than it was 18 years
ago. "Alienation" and "hopelessness" are no longer Central European words; they are very American. This broad, and often near-
nihilistic, cultural apathy and despair is relatively new on the
American (and also on the British) scene. Its existence suggests
that, despite the errors of Orwell's visions, the nightmare quality
of "1984" continues to obsess our imagination, and not merely
as the sickly titillation of a horror story. It haunts millions who
fear that life may become an Orwellian nightmare even without
the political tyranny that Orwell had predicted.

"It is by his political writings," Bertrand Russell once wrote, "that Orwell will be remembered." If this is so—and at this moment, halfway to 1984, it still seems so—he will be remembered for the wrong reasons, and one can only hope that the slow corrective tides of public opinion in the long run will redress the balance.

Orwell was not so much concerned with the degeneration of justice as with the degeneration of truth. For Orwell, both in the beginning and in the end was The Word. This is true of "1984," too, which had three levels. On the top level there is the "plot," the love affair of Winston and Julia, which is really flat and inconsequential. On the second level there is the political vision which, as we have seen, sometimes holds up, sometimes not. It is the third level, of what is happening to words and to print, to speech and to truth in 1984, which agitated Orwell the most. Indeed, this spare and economical writer chose to end the novel "1984" by adding an appendix on "The Principles of Newspeak." Orwell was frightened less by the prospects of censorship than by the potential falsification of history, and by the mechanization of speech.

The first of these protracted practices would mean that the only possible basis for a comparison with conditions other than the present would disappear; the second, that the degeneration of traditional language would lead to a new kind of mechanical talk and print which would destroy the meaning of private communications between persons. This prospect haunted Orwell throughout the last 12 years of his life. Some of his best essays dealt with this theme of falsifications of truth—even more than totalitarianism, this was his main concern. As long as people can talk to one another meaningfully, as long as they have private beliefs, as long as people retain some of the qualities of Winston Smith's mother (she had not been an "unusual woman, still less an intelligent one; and yet she had possessed a kind of nobility, a kind of purity, simply because the standards she obeyed were private ones. Her feelings were her own, and could not be altered from the outside . . ."), tyranny was vulnerable; it could not become total.

Orwell was wrong in believing that the development of science was incompatible with totalitarianism (by 1984, "science, in the old sense, has almost ceased to exist. In Newspeak there is no word for science"). As we have seen, he foresaw a decay of technology ("the fields are cultivated by horse-ploughs while books are written by machinery"). This is not what has happened; now, halfway to 1984, the fields are cultivated by bulldozers while books are written by machine-men. But Orwell was right in drawing attention to Doublethink, "the power of holding two contradictory beliefs in one's mind simultaneously, and accepting both of them," and to the desperate prospects of Doubletalk, of the degeneration of standards of language through varieties of super-modern jargon, practiced by political pitchmen as well as by professional intellectuals. There is reason to believe that, were he alive today, Orwell would have modified his views on the nature of the totalitarian menace; and that, at the same time, he would be appalled by many of the present standards and practices in mass communications, literature and publishing, even in the West, and perhaps especially in the United States.

In short, the 1984 that we ought to fear is now, in 1966, different from the 1948 version. Politically speaking, Tocqueville saw further in the eighteen-thirties than Orwell in the nineteen-forties. The despotism which democratic nations had to fear, Tocqueville wrote, would be different from tyranny: "It would be more extensive and more mild; it would degrade men without tormenting them. . . . The same principle of equality which facilitates despotism tempers its rigor." In an eloquent passage Tocqueville described some of the features of such a society: Above the milling crowds "stands an immense and tutelary power, which takes upon itself alone to secure their gratifications and to watch over their fate. That power is absolute, minute, regular, provident and mild. . . ." But when such a government, no matter how provident and mild, becomes omnipotent, "what remains but to spare [people] all the care of thinking and all the trouble of living?"

Orwell's writing is as timely as Tocqueville's not when he is concerned with forms of polity but when he is concerned with evil communication. In this regard the motives of this English

Socialist were not at all different from the noble exhortation with which Tocqueville closed one of his chapters in "Democracy in America": "Let us, then, look forward to the future with that salutary fear which makes men keep watch and ward for freedom, not with that faint and idle terror which depresses and enervates the heart." Present and future readers of "1984" may well keep this distinction in mind.

They Live in the Year 2000
by William H. Honan

MEMBERS ATTENDING the annual convention of the Life Insurance Agency Management Association at the Palmer House in Chicago not long ago were surprised to find a man named Frederik Pohl on the speakers' platform. Mr. Pohl did not pretend to know Aetna from Allstate; Prudential from Continental. A spare-framed, somewhat saturnine individual in his late 40's, he explained that he is, by trade, the editor of Galaxy magazine, and that he has been writing science-fiction stories and novels since before Buck Rogers could twirl a ray gun. He had been retained by the association, he said, as a "blue skyer," to look into the future and describe what lies ahead for the insurance business. Calmly, reasonably, even a little dryly, he proceeded to do just that—and raised a few eyebrows in the process.

Such assignments have come along with increasing frequency for Pohl during the last year or so, since it has become fashionable for businessmen to retain science-fiction writers for much the same reason that medieval monarchs used to like having a court astrologer around. Americans, it seems, have become "future-oriented," as one sociologist recently expressed it. In addition to science-fiction writers, a wide circle of professional people—"futurists," one might call them—are now busy catering to, or coping with, this new taste for tomorrow.

Some futurists, like Pohl, are unrestrained high-flyers who continually risk getting star dust in their eyes. Others—those, for example, who work for the Rand Corporation or for Resources for the Future, Inc.—are more earth-bound, limiting their forecasts to cautious extrapolations of present trends. And still other futurists—academicians especially—navigate somewhere between heaven and earth, trying only an occasional swan dive off the end of their projections.

Talking to the insurance men at the Palmer House, Pohl demonstrated in short order why he is regarded as a futurist of the wild and woolly sort. "You have expanded your market about as far as possible through selling pension plans, annuities and the like," he told his audience, "and now you ought to concentrate on your $30-*trillion* market of the future." What Pohl had in mind, he went on to explain, was cryonics, which means the "scientific" freezing of a body immediately after death, followed by storage and then revival as soon as a remedy for the cause of death has been discovered. Insurance companies of the future, he speculated—he had, as it happened, just tossed off a science-fiction novel on the subject—will be writing policies to cover the storage cost much as one today can buy "perpetual care" in a mausoleum.

There are questions, of course, that arise in such speculation. What, for example, of overpopulation? And what of the legalities? Would a marriage survive such premeditated resuscitation? Would a will become effective upon the event of freezing, or would someone's benefactor really be able to "take it with him"? Pohl asked his audience to consider these matters and told them that at a cost of $10,000 per "burial" he estimated the potential world market for such policies is $30-trillion. Pohl was not, of course, the originator of cryonics—the name is derived from cryobiology, which is the study of life at extremely low temperatures. He was embroidering on the theories of Robert C. W. Ettinger, a Highland Park (Mich.) Junior College physics teacher.

The insurance men Pohl talked to that afternoon may have

done no more at the time than exchange amused glances, but they must have come up with a start a few weeks later when news was published of the world's first cryo-customer—the body of a 73-year-old California professor of psychology, a cancer victim, which now lies in a freezer in Phoenix, Ariz. The cost, according to the Cryonics Society of California, was remarkably close to the $10,000 which Pohl had forecast.

Another of Pohl's recent clients was the Dow Corning Corporation of Midland, Mich., a maker of silicone products. Pohl was silent throughout most of the three-day management-planning session, pondering the fact that medical-grade silicone rubber has been used in the De Bakey and Kantrowitz artificial heart-assist devices, as well as other subdermal implants. On the last day of the meeting, he spoke. He said he foresaw for Dow Corning not only a vast future market in artificial organs, but an even greater potential market in cosmetics. Liquid silicone, Pohl predicted, will be injected into the body, to swell curves or wipe out wrinkles, with ever-increasing frequency. (Among those who first availed themselves of the benefits of such injection were San Francisco's topless go-go dancers.) "Silicone injection apparatus," Pohl told the executives, "is going to be as common in a girl's make-up kit as the eyebrow pencil and lipstick are today." Then he added with a smile: "You take it from there."

"I predict the future for a business or an industry," Pohl explains, rocking back in a swivel chair in the cluttered office-warehouse of the Galaxy Publishing Corporation on lower Hudson Street, "by using the science-fiction writer's habit of model-making. That is, for every story or novel we write we make a paradigm—a model—of what society will look like in the future. We try to work out the political structure, the monetary system, consider what will have happened to religion, technology, education, family relationships and so forth, and then also—equally important—take into account the factors that will not change. Then we determine the interrelationships of both the variable and the constant factors and in that way construct a total society."

Pohl grants that the businessmen he talks to do not necessarily *do* anything as a direct result of his forecasts. "The interesting thing," he goes on to say, "is that they even want to listen. It's all happened rather suddenly. The reason, I think, is that business-men have begun to recognize the accelerated rate of technological change of the era in which they live, and they're a little unnerved by it. They turn to science-fiction writers because they know this is something we've been coping with for years. They turn to us because they know we were seriously investigating the con-sequences of atomic energy, space flight and even air pollution long before these things actually came to pass."

They also turn to another futurist—one considerably more cautious than Pohl, yet in his own way equally devoted to the world of tomorrow—named Harvey S. Perloff. Dr. Perloff is a tall, affable executive of Resources for the Future, Inc., a non-profit corporation supported by the Ford Foundation that makes environment studies for business, academic and government plan-ners. (Dr. Perloff took his Ph.D. at Harvard and has served as a Presidential adviser and consultant to half a dozen foreign governments.)

In contrast to Pohl's elaborate models of complete future so-cieties, Dr. Perloff makes narrow projections based on data gath-ered from the past, on any discernible trends he can find and on whatever reasonable judgments he can bring to bear on a given extrapolation. His sensitivity to trends, for example, has lately caused him to take a fresh look at such concepts as: What is a "resource"? "In the past," he explains, "we thought of a resource as a needed raw material such as iron ore, coal or something of that nature. Then if we wanted to predict where the great industrial centers would be located, we looked for the avail-ability of those materials. Today, however, we find industry mov-ing to California, Florida, Texas, New Mexico, Arizona and some people conclude that industry has broken away from its resources. But I don't agree. It's just that today, and increasingly in the future, industry is seeking out a wholly different type of resource. These are the amenity resources—good climate, beaches, recrea-tional facilities—and social-cultural or social-*ambiance* resources such as the quality of schools, social atmosphere and the type of culture represented by a community.

"You see, getting power and raw materials today is really no problem for a great many industries. What's in short supply is talent. And so industries have got to establish themselves where both the amenity resources and the social-cultural resources are attractive to talented people who can choose where they're going to live and work.

"With this new concept of resources in mind," Dr. Perloff continues, strolling over to a large wall map of the United States and thrusting a finger at Biloxi, Miss., "I'd say one of the towns along the Mississippi Gulf Coast here would be a great place to invest some money if you could afford to wait a while. Right now, the racial environment in these towns makes them not very attractive to many skillful people—particularly those associated with the glamour industries. Nobody is going to set up an R.-and-D. firm or a high-level electrical equipment plant there today, because he'd never get the scientists and the engineers to come and raise their children in such a community. But once the racial environment, which is after all one of the social-cultural resources, becomes moderate—as, for example, in Atlanta or Miami—and these towns can begin to capitalize on their wonderful climate and beaches and so forth they will bloom economically.

"By the same token"—Dr. Perloff gestures to the other side of the wall map—"the abundance of these new amenity resources and social-cultural resources, especially education, is one of the main causes of California's spectacular growth in the last few years. Just now there's a danger that Governor Reagan's handling of the University of California may reverse that trend. I would guess that if the trouble continues, and the top scholars start an exodus from the state, the aerospace people will follow, and the R.-and-D. and other glamour industries will not be far behind. The university, joined together with the state's magnificent physical resources, plays very much the same role in attracting people from all over today as did California's gold mines in an earlier time."

Another futurist, who plies a middle course between the giddy flights of Pohl and the cautious calculations of Dr. Perloff, is Daniel Bell, a compact, fast-speaking Columbia University sociologist (currently on loan to the University of Chicago) and chairman of the American Academy of Arts and Sciences' Commission on the Year 2000. Bell has been enchanted by the future for years, and has been peppering the not-so-mass media with a series of vigorous, brilliant and always controversial comments on the subject for more than a decade. His interest, he says, stems partly from his early fascination with Marxism ("The Marxist always asks what comes *after* capitalism"), partly from his academic specialty ("There's a built-in notion in sociology that society is not permanent, that things will change"), and partly from his 10-year stretch as labor editor of Fortune magazine ("It's not unusual at all for businessmen to plan 25 years ahead").

Bell gives in to his urge to play the prophet not without a sense of irony. In 1957, for example, when he addressed a conference of Kremlinologists in Oxford, with a speech subtitled "The Prediction of Soviet Behavior"—which later became a chapter in "The End of Ideology," the book that solidly established his reputation—Bell could not resist mocking himself as a sort of poor man's Pirandello gamely, but also a little wryly, "pursuing the illusion of reality." Three years later he made another stab at prophecy in a much more straightfaced essay for Daedalus, the journal of the American Academy, which he called "Twelve Modes of Prediction." Even there, however, he noted with some chagrin that his preoccupation with the future was perhaps "the modern *hubris*." And, in October 1965, when he opened the first meeting of the Commission on the Year 2000, Bell felt constrained to say that while the commission's undertaking was one of "rather extraordinary imagination and daring," there was also present "a touch of preposterousness."

But, irony aside, Bell is fundamentally convinced that the future, or at least what he calls "alternative futures," is there for those brave enough to reach out and grasp it. Acting on this assumption, he assembled on his commission of prognosticators 27 of the most brilliant economists, political scientists, biologists, sociologists, historians, legal scholars, administrators and statisticians he could find, and asked them to help him dope out what the world might look like 33 years hence, and to suggest as well what ought to be done to prepare for, or to ward off, those eventualities. "Is it not a fundamental responsibility for a society which has become as interdependent as this one," he asked his colleagues at their first meeting, "to try some form of anticipation, some form of thinking about the future?"

His challenge was evidently greeted with exuberant yeas, for the Commission on the Year 2000 has now held two all-weekend sessions and has mimeographed a stack of reports which include such documents as Prof. Harry Kalven's "The Problem of Privacy in the Year 2000," which contemplates a decline in privacy "to the point where an interest in solitude will be regarded as eccentric and queer and something like homosexuality," and Prof. Howard Becker's "Some Notes on Hip and Square in the Year 2000," which envisions the possibility of "an apocalyptic confrontation" between hipsters (beatniks in coalition with liberal ministers) on the one hand and squaresters (the old bourgeoisie allied with student activists) on the other.

Bell and most of his fellow commissioners consider that there are two basic sorts of prediction—first, the forecasting of events and inventions which Bell regards as "very difficult," if not impossible, and second, sociological predictions which he thinks of as "remarkably easy." In the first category, for example, Bell points to the fact that before the First World War virtually all the experts believed that if hostilities were to break out, Germany could last only three months because the inevitable Allied blockade would cut off its supply of Chilean nitrates, which were indispensable in the manufacture of explosives. Allied strategy was mapped accordingly. At almost exactly the moment at which the war finally erupted in 1914, however, Fritz Haber and Karl Bosch mastered the industrial synthesis of ammonia, and suddenly the Kaiser's war machine was no more dependent on Chilean nitrates than on Peruvian arrowheads. No one, says Bell, could have foreseen that.

Political events are also largely outside the realm of prophecy, Bell believes, because of what he calls "Brzezinski's Law." A couple of years ago Bell happened to see on television his old friend Zbigniew Brzezinski, the director of the Research Institute on Communist Affairs at Columbia, and a fellow 2000 commissioner. Brzezinski was being twitted for not having been able to foresee the ouster of Premier Nikita Khrushchev, and he finally defended himself with a Slavic shrug of despair and said: "If Khrushchev couldn't predict his downfall, how would you expect *me* to do it?" Bell's point (and Brzezinski's) is that when given imperfect, or uncertain, data—as is almost always the case in politics, even if they come from the participants themselves—a basis for prediction does not exist.

In contrast to such uncrackable nuts, sociology, in Bell's view, is the prophet's dreamworld. A great many of the predictions of the past generation of social scientists, he points out, have come true in recent years. Many of the features of American society today, for example, ranging from medical and educational needs to the problems of poverty and racial conflict, were predicted with striking accuracy, he says, by the group of scholars on President Hoover's Research Committee on Social Trends in reports published in 1933 entitled "Recent Social Trends."

Buoyed by such successes of the past, Bell, summarizing his own as well as his commissioners' prognostications, ventures into the future with a set of "baselines," as he calls them, which help to delineate problems relevant to the present, and therefore to the future: "Without in any way arguing that this is actually the range of all things we can speculate about, here are just a few notions of what, presumably, the world or America may be like.

"In the year 2000, world population will be 5.1 billion, 55 per cent over that of today. Our national income will have more than doubled. There will be a high extension of the G.N.P. so that leisure and the use of time will become matters of concern. With income so high, people will be able to afford a variety of public services and rehabilitation schemes—for cities, housing, air, water, etc.—about which they now grumble about paying for. There will also be a diminution of conflicts such as those that now exist between labor and capital. With plenty of money, people won't mind giving up marginal portions of what they're getting.

"Men will live longer, and the life cycle will become more and more of a problem as people are not simply pursuing one career, but going through different career cycles. There will be a diffusion of upper-middle-class attitudes—the desire to take part, to help decide, to be heard. The lessened influence of the family upon various groups in the society may be a problem. People will be more mobile and live in a more crowded world. Interaction with a remote world will be common. A universal language will have evolved through automated communication. There will be greater consciousness of mass communication. The problem of indecisiveness about what to educate for will increase, largely because people will be unaware of, or insecure about, their ability as a result of the many new tasks confronting them."

Turning to still darker possibilities, Bell continues: "There is likely to be a breakdown of the service industries, what with the enormous maintenance load; also, a serious problem with the socialization of the adolescent, and, even more immediately, a class society in America which has become color society. That is, inasmuch as education is the ladder of social mobility, the disproportionate dropout rate of Negroes in school at present marks them for the bottom positions in the society of the future."

One of the fundamental reasons for making such forecasts as this, Bell explains, is precisely so that the undesirable ones need not come true. "The function of prediction," he wrote a few years ago, "is not, as is often stated, to aid social control, but to widen the spheres of moral choice." At the first meeting of the Commission on the Year 2000, he told participants: "We want to let people choose if they can, and we want to get mechanisms which allow us to intervene, and to change to a different path. This is the thing we are all trying to talk about here."

Daniel P. Moynihan, director of the Harvard-M.I.T. Joint Center for Urban Studies, said at the same meeting: "It seems to me that one of the most important events of the future is the development of this forecasting technique."

Given such approaches as those of Pohl, Perloff and Bell, how much cause is there for real optimism about prognostication of the future? Unfortunately, one is not exactly transported with hope by the sallying forth of even our bravest and most resourceful scouts. Mr. Pohl's speculations are certainly intriguing. However, one is forced to agree with Daniel Bell that *technological* forecasting, at least, is beyond the realm of possibility. To be sure, there are those who have succeeded. Much is always made, for example, of the "prophetic" novels of Jules Verne and H. G. Wells. Less famous, but equally amazing, are Anatole France's description of the destruction of Paris with a "radium" bomb in "Penguin Island" (1908); Samuel Johnson's comment in a letter written 80 years before the discovery of microbes that such invisible creatures would be found to be the cause of dystentery, and finally Swift's prophecy in "Gulliver's Travels" that Mars would be found to have two moons.

Incredible as such forecasts may seem, they are essentially "broken-clock predictions." "Even a broken clock," the old French proverb has it, "is right twice a day." Considering the vast number of speculations that have been made, it is not improbable that some turned out to be right, and a few stunningly so. Far more representative of technological predictions, however, are the blunders—the great Goethe's forecast that the microscope and the telescope would be "of no use whatsoever"; Alfred Nobel's belief that his invention of dynamite would make wars "altogether impossible"; Simon Newcomb's "complete proof" that the flying machine was not feasible; a British parliamentary commission's declaration in 1878 that Edison's electric light bulb was "unworthy of the attention of practical or scientific men"; Einstein's view, as late as 1933, that the unleashing of nuclear energy was not worth the effort; and so on, ad infinitum. "The forecasting of invention," as one member of the Commission of the Year 2000 put it, "is itself invention."

The possibility of sociological prediction also bears close scrutiny. The "Recent Social Trends" series, for instance, which Bell cites as an illustration of the soundness of such leaps into the future may just as well be used to demonstrate their recklessness. Take the population essay which was contributed to the series by Warren S. Thompson and P. K. Whelpton, two of the leading demographers of President Hoover's day. Their projection was that "the population [of the United States] will reach its greatest size—approximately 146 million—between 1965 and 1970 and will subsequently decline." We are now just about midway into the period covered by that estimate, and the population of the country has not only already exceeded their figure by 52 million, or 36 per cent, but, far from going into decline, is now expected, according to the most recent data from the Census Bureau, to soar to 338 million by the year 2000, which is a good deal more than twice what the Hoover Committee thought likely.

This is not to say that all sociological prediction goes out the window. Other forecasts of the Hoover Committee were indeed close to their mark, as Bell points out. Nevertheless, if the prediction of such basic sociological data as population growth—data upon which numerous other predictions must rest—is to be this hit-or-miss, the failure is substantial enough to raise the question of whether forecasting of this sort may not actually close more doors than it opens, or perhaps open all the wrong doors while closing the right ones—contrary to what the Commission on the Year 2000 so laudably intends.

Another danger inherent in what the futurists are attempting—and one that Bell himself finds disturbing—is that speculation about the future may induce a sort of moral simple-mindedness. The leap into the future is too often a bolt away from the stubborn, the grubby, the unfathomable and the infuriating complexities of the present into a sort of carefree playland in which the shorter work week appears to promise more water-skiing (but not more alcoholism) and a helicopter is pictured on every suburban rooftop (rather than in more Vietnams). This dimple-cheekedness was particularly evident in an anniversary issue of Look magazine a while back in which a host of celebrities were invited to unveil their private visions of the blessings that lie ahead. The contribution from Mike Nichols and Elaine May reads as follows: "By 1987, the world will be at peace. Those forces that have been so expensively and successfully developed for destructive purposes will have been converted toward peaceful and constructive ends. There will be no more disease, no more tyranny, no more graft, no more suspicion, no more poverty, no more war—and we are the Andrews Sisters."

A final caveat is that even if it were possible for the futurists to anticipate history with absolute clairvoyance, who would listen? Is society rational enough to take advantage of foreknowledge, or is man innately and intractably stuck on muddling through? It gives one pause to recall that the coming of the present air-pollution crisis—just to light upon a handy illustration—was predicted by any number of Cassandras as long as 10 years ago. Yet it was politically impossible to get anything done about the problem until after last Thanksgiving, when New Yorkers in Times Square actually began to feel their eyes smarting from sulphur dioxide.

To this argument, the futurists reply that if the present political structure is indeed not capable of assimilating and acting upon foreknowledge, the need to do so is now so critical that it is incumbent upon this generation of Americans to create new political techniques and institutions whereby society *will become* responsive. The technocracy of tomorrow, they say, must necessarily function something in the way that the Department of Defense is supposed to be working today, keeping its planning a generation ahead of operations, preparing simultaneously for a whole range of future possibilities.

At this point in the discussion, it is perhaps worthwhile to observe that, in the final analysis, what is needed in our attempts to cope with the future is a balance between the impetuous desire to know and a wise respect for the unknowable. This balance, and the means of maintaining it, was beautifully described by a futurist of no mean order, the late J. Robert Oppenheimer, nearly two decades ago. "When the time is run, and the future become history," he said, "it will be clear how little of it we today foresaw or could foresee. How then can we preserve hope and sensitiveness which could enable us to take advantage of all that it has in store? Our problem is not only to face the somber and the grim elements of the future, but to keep them from obscuring it. . . .

"[The means] of doing justice to the implicit, the imponderable, and the unknown," Oppenheimer went on to say, ". . . is sometimes called style. It is style which makes it possible to act effectively, but not absolutely; it is style which, in the domain of foreign policy, enables us to find a harmony between the pursuit of ends essential to us and the regard for the views, the sensibilities, the aspirations of those to whom the problem may appear in another light; it is style which is the deference that action pays to uncertainty; it is, above all, style through which power defers to reason."

Third Great Revolution of Mankind

by Charles Frankel

PROBABLY NO EVENT in recent years has had a greater impact on the American mood than the Soviet Union's success in launching artificial satellites. The sputniks stirred justified anxieties about America's position in the cold war and upset some of our tenderest assumptions about our situation. But more than this, the sputniks—and now our own Explorer—have turned loose some bewildering predictions about the future.

Trips to the moon and twenty-minute jaunts to Moscow, we are told, are on the agenda of human "progress"—though no one has yet said how to make the moon or Moscow more attractive places in which to land. It is even whispered that mankind may lose one of its oldest topics of conversation because it will be possible to control the weather.

But while predictions about man's future career in outer space have an undeniable interest, the satellites, it may be suspected, also have a significance that is a bit closer to earth and closer to home. For they are not merely events in an armaments race or science-fiction stories come true. They are peculiarly dramatic

From the *New York Times Magazine,* February 9, 1958, copyright © 1958 by The New York Times Company.

symbols of what has been going on backstage, East and West, during the last fifteen years—a sudden extension of scientific intelligence and technical resourcefulness which represents an extraordinary spurt of human intellect and power. And this spurt of the human mind has social and moral, as well as technological, consequences. The sputniks are signs in the skies that the normal human scene is changing in some of its fundamental characteristics and that we are living in the midst of a fundamental revolution in human affairs.

Some 25,000 years ago an "Agricultural Revolution" took place which changed man from a nomadic hunter and berry picker into a deliberate cultivator of his food supply. In the latter half of the eighteenth century an "Industrial Revolution" began, with results which we have not yet fully absorbed. Both these revolutions began as changes in the ideas and tools men had used to adjust themselves to nature. They ended by changing men's relations to each other, their moral and political outlooks, and the very substance of the things they thought worth seeking in life. It is easy to overestimate the significance of events that happen in one's own day. But the revolution that has taken shape in the last fifteen years must be put in company like this to be seen in its proper perspective.

Indeed, in the natural energies it has released, and in the speed with which it has done so, the present shift in the relation of man to his environment dwarfs either of its predecessors. It is impossible to believe that its other consequences will not eventually be as great. Now that we have had a chance to absorb the first impact of the sputniks, it may be worth while to sit back and reflect on some of the long-range social issues of which the sputniks are a portent.

We can already see some of the more obvious issues. War, for example, has changed its character and has lost one of its traditional functions in international affairs. Leaving all issues of morality aside, large-scale war can no longer be used, as it has sometimes been used in the past, as an intelligent instrument even of national selfishness. While the danger of all-out nuclear war has not substantially receded, such war can only be an instrument of utter desperation.

Similarly, the problem posed by the expansion of population in the world promises to become even more acute as a result of the advances in medicine and technology that are almost surely in prospect. Through most of its history the human race has had to struggle to keep its numbers from declining. But our very success in improving the basic conditions of human existence now threatens to turn back upon us and to lead to incalculable human suffering unless organized measures are taken to control the birth rate.

But war and the growth of the world's population are relatively familiar problems, even though the penalty for failing to solve them has suddenly become astronomically high. The present revolution in human affairs is likely to bring other changes, however, to which somewhat less attention has been paid. And not the least important is a possible change in the way in which human work will be organized with the advent of new industrial processes such as automation.

One possible consequence of automation, for example, is a sharp increase in the ratio of skilled workers to unskilled workers. This means a host of new issues for industrial unions, and new problems for both labor leaders and industrial managers. Of equal significance is the possible impact of the automatic factory on the way in which the working day may be arranged. As the British engineer Landon Goodman has pointed out, the cost of introducing automation may be so high in many cases that it will be uneconomical to operate a plant only eight hours a day. If many industrial plants are going to find it necessary to operate around the clock, obvious consequences will follow for everything from the nature of home life to the way in which cities are organized. Even the old phrase, "as different as night and day," is likely to lose some of its force.

The new way in which work may be organized also affects the attitudes that men may take toward other parts of life. Most of the work that men have had to do in history has been disagreeable; most of the leisure that men have had has been the pre-rogative of the few. This fact has colored much of our thinking about the way in which life ought to be lived. The democratically-minded have been suspicious of what is "useless"; the aristo-cratically-minded have regarded the useful as just a bit vulgar. But if leisure becomes everybody's prerogative (and problem), and if automation can be used to make human work less routine and to give more workers the opportunity to exercise their indi-vidual skills and discretion, the sharp division between work and leisure will make even less sense than it makes today. The effects will be felt, to take only one example, in our ideals of "liberal" education, which are still primarily leisure-oriented, and in our conception of "vocational" education, which is already anachro-nistic in its view of what ordinary people need to be "prepared for life."

But the new processes of industrial production are parts of a larger trend which has even deeper implications of its own. Dur-ing most of the past, developments in technology were largely independent of developments in pure scientific research. To some extent this remained true even in the nineteenth century. But technology has now become almost entirely the child of funda-mental theoretical inquiry. This means that we can count in the future on a steady process of technological innovation, and at a steadily more rapid pace.

We come at this point to perhaps the profoundest consequence of the present revolution in human affairs. It is the simple change in the tempo of change. For nothing cuts more quickly or deeply into a society's way of doing things than changes in its technology.

This quickened tempo represents an unprecedented challenge to the human ability to adjust to social change. It took man roughly 475,000 years to arrive at the Agricultural Revolution. It re-quired another 25,000 years to come to the Industrial Revolution. We have arrived at the "Space Age" in a hundred and fifty years —and while we do not know where we go from here, we can be sure that we shall go there fast. Our expectations of change, and the ability of our nervous systems or our social systems to with-stand the shock of change, have been formed in the long experience of the race. And this experience, even in the nineteenth century, has not prepared us for the pace of events that lie ahead.

Such an extraordinary change in the basic tempo of human history means that new and deliberate efforts will be needed to control the processes of social change. As the last hundred years of Western history demonstrate, men can learn to change at a much quicker pace than before. But as these same years also suggest, there are limits, and it is difficult to imagine a day when it will not take time for men to adjust to new conditions, to learn new skills and habits, and to get over the nostalgia and resentments that come when old and familiar things are destroyed. There is a conservative in every man and, in the world into which we are moving, he is going to get a harder workout than ever before.

Accordingly, if the things we cherish from the past are not going to be carelessly destroyed, and if the best possibilities of the future are going to be realized, it seems probable that we shall have to have institutions that have been deliberately set up to exercise long-range social forethought. A steady process of technological innovation, for example, can mean recurrent crises of technological unemployment. If this is not to happen, institutions will have to exist to envisage the new skills that will be needed, to undertake the continuing task of retraining workers, and to control the pace at which new techniques are introduced so that we can make a sensible adjustment to them. Given the pace and magnitude of the technical changes that are in prospect, we cannot count on the market place and the price system to do this job alone. Technological innovation means social change; and there is no more reason to introduce such innovations, letting the chips fall where they may, than there is to introduce a new and powerful drug on the market without first making it meet the test of medical examination and control.

The need to exercise more deliberate control over the processes of social change raises, of course, a fundamental issue. It is the issue of freedom and regimentation, the question of the tension between personal liberty and initiative on the one hand, and the obvious and growing necessity, on the other, for an ever larger degree of social organization. This has been the central issue in industrial society for more than a century. In the world which the present revolution is forming it' will be equally decisive. But the terms we have habitually employed in trying to solve it will almost certainly need to be revised.

The dangers that a larger degree of social organization can bring are obvious. It can mean, at the very least, a multiplication of the nuisances that exhaust individual energy—administrative forms to fill out, incessant committees, petty bureaucratic tyrannies. It can mean that power will steadily pile up at the center until it is impossible to place an effective check upon it. Perhaps worst of all, our ideals and attitudes can change.

Under the pressure of the need for organization, we can come slowly but painlessly to like the standard and the impersonal, and to prefer the man who fits the system to the man who is difficult to harness. If this happens, we can lose liberties we now cherish and never notice or regret the loss.

But the individual can be crushed just as easily in a subway rush as on an assembly line. When there is more traffic on the streets, more controls are necessary. And when these controls are inadequate or break down, the individual has less freedom to go where he wants to go, not more freedom. The question, in brief, is not whether we shall have a larger amount of conscious social organization or not, but what kind of social organization we shall have. It can be centralized or decentralized; it can be broken up into small units or cover only very large ones; it can concentrate authority up at the top or spread considerable authority down to lower levels. Most important of all, it can have as its conscious object the cherishing of individual differences and the promotion of individual talents. The dangers of a larger degree of social organization are unmistakable. But disorganization will be no healthier a climate for freedom.

The problems of social organization and of controlling the results of technological innovation bring us, however, to a final problem. It is the problem of the use and abuse of science, a problem that is likely to become steadily more acute as our world becomes more steadily and obviously a creature of science.

In the future, as in the past, two extreme possibilities confront us. The first is to make an idol out of science. The second is to denigrate its importance on grounds of fixed moral or religious principle.

Illustrations of the tendency to make an idol out of science have suddenly blossomed all around us since the sputniks began their dance in space. With a hopefulness that is somewhat frightening, both scientists and laymen have predicted, for example, that science will soon be able to change human emotions and desires by biochemical means. Others have talked about the coming mastery of the principles of "group dynamics" and the ability that science will give us to choose the right leaders and to get people to work together harmoniously.

Such predictions represent revivals of Plato's ancient dream that, if philosophers became kings, man's political and moral troubles would come to an end. And they rest on precisely the same combination of political innocence and moral arrogance. There is, unhappily, no guarantee that those who dispense the pills that are going to change our desires and emotions will themselves have the right desires and emotions.

But it is a grave mistake to dismiss science as useless in solving moral and political problems. Objective knowledge of the conditions and consequences of our personal desires or our social institutions does help us to realize the actual nature of the ends we choose to pursue; and in this way we can frequently come to choose our ends and ideals more intelligently.

Even more than in the past, the world which the present revolution is creating will be one in which a process of steady reexamination of existing institutions will be a condition, not simply of a decent life, but in all probability of survival as well. Those who take fixed positions in such a world, and who deny the usefulness of scientific knowledge in resolving moral and political dilemmas, will be pleading merely for the rule of dogma or of their own private intuitions. It is unreasonable and unattractive to think that the society of the future should be ruled by a scientific élite masquerading as moral experts. It is equally unattractive to think that it should be ruled by those who make a principle of ignorance, and whose claim to be moral experts rests on their sense of superiority to the processes of sober scientific inquiry.

The attitudes that are aroused by science suggest, indeed, the fundamental educational problem which the present revolution puts before us. This revolution is at bottom the product of ideas and modes of thought which have remained closed secrets to most of the best educated men and women today. As a result, in what is alleged to be the most "scientific" of ages, science has the quality of magic for the popular mind. But while the problem is critical, it is not insurmountable.

The difficulties of acquainting college-trained men and women with the fundamental methods and ideals of contemporary science have been greatly exaggerated. It requires, of course, the imparting of factual information; but it requires, even more, the training of the imagination of the ordinary, educated layman so that he can grasp the general character of scientific problems even if he does not understand all their details, and can appreciate the kind of triumph which the solution of such problems represents.

Such an imaginative grasp of science, which would allow more members of modern culture to share vicariously in the most majestic achievements of their civilization, is possible for a great many more than now have it. And even more than the problem of training more scientists and engineers, this is the fundamental problem of education in the sciences.

As one looks ahead to the unfamiliar world that is emerging, it is possible, of course, to feel overwhelmed by this educational problem or by the other problems which this world puts before us. One can try to retreat from the unfamiliar, either by laughing off artificial satellites as mere basketballs in outer space or by concentrating almost hysterically on just one short-range issue— the military struggle for outer space—to the neglect of all the other issues that artificial satellites dramatize. And one can also take an apocalyptic attitude, and assume that the unfamiliar world that is emerging is also going to be absolutely unrecognizable, whether for the better or for the worse. But human traits like envy, malice and egoism are likely to remain with us no matter what moral medicines the druggist of the future has on his shelves. And once the initial thrill wears off, most honeymooners are probably going to prefer the moon overhead rather than underfoot.

But if utopia is not around the corner, neither is it inevitable that our powers are unequal to the problems that are appearing. In an age whose problems are almost all signs of mounting human powers, this would be a strange moral to draw. Man is now making his own stars and setting his own impress on the solar system. If these stars are as yet minuscule and only a very little way out in space, they still represent something of an achievement for a creature who is built rather close to the ground. The world that is taking shape can preserve old joys and it can also contain many new ones. The scientific imagination of the twentieth century has shown remarkable flexibility and daring. There is no reason in the nature of things why our social imagination cannot show some of the same qualities, or why it cannot escape, as modern science has, from the backyard of its old commonplaces and dogmas. If it did, its achievements could be even greater than the shooting of satellites into the sky.

Not the Age of Atoms but of Welfare for All

by Arnold J. Toynbee

IN THE CONTEMPORARY world—in whatever age or century one happens to be living—religious and political differences between various sections of the living generation are apt to seem absolute and ultimate. For instance, our seventeenth century ancestors in Western Christendom could not conceive of any greater gulf than that which seemed in their day to be fixed between the Catholic and Protestant varieties of Western Christianity. By contrast, we their descendants, looking back on them and their conflicts in the perspective of three centuries of history, are far more conscious of the gulf between our own age and the world of the seventeenth century than we are of the domestic divisions within that seventeenth century world. To our eyes, those Protestants and Catholics are all alike, seventeenth-century-minded people first and foremost; and it needs some effort of discrimination on our part to appreciate the nice distinction between the warring religious camps.

In the light of this historical precedent—and there are countless others that would have served equally well to illustrate our

From the *New York Times Magazine,* October 21, 1951, copyright ©
1951 by The New York Times Company.

point—we may be sure that, 300 years from today, our own descendants will be much more alive to the common features of the twentieth century world—and especially those common features that seem to them distinctive—than they will be to the current differences that mean so much at the moment to all members of the living generation, in whatever continent or camp we may happen to have been born.

Can we guess what the outstanding feature of our twentieth century will appear to be in the perspective of 300 years? No doubt we shall not all guess alike. Some of us will guess that the present age will be looked back upon as the age of scientific discovery. Others will expect to see it branded as the age in which Fascist and Communist apostates from a Christian civilization harnessed science to the service of a neo-barbarism. My own guess is that our age will be remembered chiefly neither for its horrifying crimes nor for its astonishing inventions, but for its having been the first age since the dawn of civilization, some five or six thousand years back, in which people dared to think it practicable to make the benefits of civilization available for the whole human race.

By comparison with the significance of this common twentieth century new ideal, the differences between the conflicting ideologies will—so I should guess—come to look both less important and less interesting than will be easily credible to anyone alive today. In the easy wisdom that comes after the event, our successors will, perhaps, be able to pronounce that this or that policy for achieving a common twentieth century ideal was more suitable than the rival policy was to the social conditions of this or that region in that antique and still unstandardized twentieth century world.

They may even judge that one twentieth century ideology was better or worse than another in some absolute moral sense. But the common features of our century will, I fancy, be the features standing out the most prominently in perspective; and, among these, the new ideal and objective of extending the benefits of civilization to the common man will in future centuries tower above the rest.

Perhaps there are two points here that are worth underlining: This vision of a good life for all is a new one, and—whatever our success or our failure may be in the attempt to translate this vision into reality—this new social objective has probably come to stay. That the ideal of welfare for all is new is surely true; for, as far as I can see, it is no older than the seventeenth century West European settlements on the east coast of North America that have grown into the United States. And it has surely come to stay with us as long, at any rate, as our new invention of applying mechanical power to technology; for this sudden vast enhancement of man's ability to make non-human nature produce what man requires from her has, for the first time in history, made the ideal of welfare for all a practical objective instead of a mere utopian dream.

So long as this aim continues to be practical politics, mankind is certain, however many times we may fail, to go on making one attempt after another to reach the goal. When once the odious inequality that has hitherto been a distinguishing mark of civilization has ceased to be taken for granted as something inevitable, it becomes inhuman to go on putting up with it—and still more inhuman to try to perpetuate this inequality deliberately.

Of course it was one particular ingredient in welfare—a spiritual ingredient which was at the heart of it—that was the objective of the first settlers from the Old World on this American coast. They were inspired to pull up their roots in the Old World and to set about the creation of a new world beyond the Atlantic by the hope of being able at this cost to purchase liberty—religious liberty above all, since they were living in the seventeenth century. But, as soon as they realized that their quest for liberty had landed them on the edge of a vast fallow but cultivable continent, the vision began to dawn upon their minds and hearts of offering the opportunity of a good life for everybody—by offering everybody the opportunity of carving a farm for himself out of a seemingly illimitable expanse of potentially arable land.

The ideal of welfare for all came into the world initially in

North America in the eighteenth and nineteenth centuries because the sudden acquisition of immense new virgin material resources here for the first time made this vision seem practical.

It is noteworthy that the North American society which first conceived of this ideal was still living in the Old World, though it had established itself on the American side of the Atlantic. It was still living in the Old World in the sense that it remained in a pre-industrial age in which the natural resources that were the material bases of civilization were the crops and cattle on which the earliest civilizations had been reared.

The original North American version of this new ideal was still an ideal of welfare for everybody in an old-fashioned agricultural society. In the civilizations of the past there had been a just sufficient stock of arable land to provide a bare subsistence for all and a good life, in addition to a subsistence, for a very small minority. It was manifest to everyone that the normal resources of an agricultural society could not maintain more than a small minority of the whole population at a level higher than that of bare subsistence.

The chance of welfare for everybody in the North American agricultural society, which came after the plow first broke the continent's virgin soil, was no more than a limited and a transitory one. Vast though the untouched reserves of arable land in North America might seem in the eighteenth and early nineteenth centuries by comparison with agricultural opportunities in an agriculturally long since congested Western Europe, the North American continent was only a small fraction of the whole habitable world, and it took little more than 100 years to bring North America's fields under cultivation.

If the new material resources required in order to make practical politics of the new ideal of welfare for all had been confined forever to new agricultural resources, the dream would soon have faded away again. After the conquest of North America by the plow, the only remaining virgin soil in the Temperate Zone was Manchuria; and after the plow's conquest of Manchuria in the early twentieth century, the future of mankind, as a whole, so far from being anything like "welfare for all," would have been something like the present as it can be seen in China or India.

The reason why "welfare for all" is still practical politics in the world at large today is because we have tapped a wholly new kind of material resource in discovering how to harness mechanical power to technology. Mankind's hope of better things lies in a permanent industrial revolution.

As a twentieth century non-American sees it, looking back on nineteenth century American history, the American outlook, like all particular outlooks, is based on a particular experience; and the particular experience that has molded current American ideals to their present shape is that experience of stumbling upon a whole continent of virgin arable soil. The ground for the American hope of providing a good life for all was expressed in the two nineteenth century American magic words, "Go West."

In a nineteenth century agricultural United States, the local and temporary existence of empty arable lands did indeed give to the weaker party in the economic arena so effective a bargaining power in his dealings with the stronger party that it was possible for the weaker party to win his fair share of welfare without its being necessary to curb the stronger party's freedom of economic action. Even under the very different American conditions of today, enough of these nineteenth century agrarian American circumstances perhaps still survive in a twentieth century industrial America for the best of both worlds to be still more or less practical politics locally in the United States.

By "the best of both worlds" I mean, of course, a maximum of opportunity for all, combined with a minimum of restriction upon a stronger and wealthier minority's freedom of action. But if this state of relative felicity is perhaps still attainable locally in the United States, it certainly is not, any more than it ever has been, practical politics in the world at large.

The outlook of the twentieth century world at large is governed, as I see it, by two facts. The first fact is that three-quarters of mankind are today still living the traditional life of an agricultural civilization in which there is no reserve of virgin soil and therefore no possibility of providing more than a tiny minority of the population with anything better than bare subsistence out of agricultural production.

But, in this old-fashioned starveling agrarian world, the Industrial Revolution has brought with it a hope for all mankind, from the prosperous American technician and farmer to the most miserable Chinese or Indian coolie, of breaking right through the iron limits to which the extension of the benefits of civilization has normally been subject in an agricultural society.

This hope is now rapidly dawning in the hearts of the depressed and ignorant peasantry that today still constitutes three-quarters of the living generation of mankind. They have begun to ask themselves how they are to attain those benefits of civilization which a mechanized technology has at last brought within the horizon of every man's hopes. But, considering the greatness of the gulf between present Asian and present American circumstances, it seems unlikely that the common Asian and American objective of extending the benefits of civilization to every man by drawing on the new resources of a mechanical technology can be attained in Asia in exactly the American way.

A common goal has to be approached along different roads by people who start their journey toward it from different quarters of the social compass. We must therefore expect to see an ideal which Americans have brought into the world being pursued by Asians and Africans on lines which, in contemporary American eyes, may, at best, look strange and, at worst, look misguided.

How is this depressed three-quarters of mankind going to set about the stupendously difficult task of gaining the benefits of civilization? Now that the hundreds of millions of peasants are aware of the relative well-being of the Western peoples, nothing is going to stop them from setting out to reach a goal which the West seems to them to have attained already. And no doubt only trial and error are going to make them aware of the difficulties in their path which are glaringly manifest to Western eyes.

For us Westerners it is easy to see that the mass of mankind today does not command those assets and advantages which have enabled a Western minority within the last two centuries to make some progress toward a wider distribution of the benefits of civilization inside the narrow circle of our Western society. Unlike nineteenth century and twentieth century America, they have no great installations of industrial plant, no human fund of widespread technical skill, no professionally competent and experienced middle class and—most serious deficiency of all—none of those Western traditions and habits of personal conduct which are the ultimate source of all the West's material success. If the mass of mankind did appreciate the seriousness of these handicaps, they might indeed be discouraged, but we can already see that this practical side of the problem is not going to be uppermost in non-Western minds.

Asia and Africa are going to make an audacious attempt to catch up with the West by a forced march, and here lies communism's opportunity in a world in which the Russian ideology of communism is competing with the Western ideal of free enterprise for Asia's and Africa's allegiance.

The present state of mind of an awakening majority of the human race in Asia and Africa is communism's opportunity because a forced march can never be made without severe regimentation and discipline. The rulers of the Soviet Union can plausibly represent to the rest of the non-Western world that their own system of totalitarian government already enables the peoples of the Soviet Union to overcome just those practical obstacles by which other non-Western peoples are faced.

This Russian claim is bound to appeal to Asians and Africans who are eager to reach the goal of welfare for all and who are also in a mood in which they will give priority to equality over liberty in a situation in which they may have to choose between the two. And this inclination in Asian and African minds to see salvation in communism is bound to be riling for Western observers.

What, then, is our Western policy to be? Being human, we

might be tempted to give way to our sense of annoyance. Why not wash our hands of this whole Asian and African business? Isn't this dream of welfare for all mankind just a folly? And, if they are hoping for salvation in communism, cannot we count on their being disillusioned sooner or later? Such a reaction on our part would be as natural as it would be rash and wrong.

It would be rash because we could not be sure that the Soviet Union might not secure a political and military hold over Asia and Africa before the process of disillusionment had time to work itself out. And it would also be wrong for us to hold aloof because we human beings are, after all, our brothers' keepers, and we cannot be indifferent to the fate of three-quarters of the human race.

In this difficult situation the supreme need of the hour is, I would suggest, an immense patience and mutual toleration. A revolutionary improvement in means of communication has suddenly brought peoples with sharply diverse traditions or civilizations into close physical contact with one another. If, in spite of our diversity, we find ourselves all alike pursuing what is ultimately a common ideal, this is something to be thankful for. If we believe in the freedom which the Pilgrim Fathers came to look for on American shores, we must believe in the right of each people to work its way toward our common objective along a freely chosen course of its own.

Acting on this belief means acting toward Asian and African countries as we are already acting toward Yugoslavia. We must be ready to work with any non-Western country which agrees with the West on the crucial political point of being determined to resist Russian attempts to dominate the world, and we must not lay it down as a condition for their receiving help from us that the people who are asking us for assistance shall pursue our common social aim of welfare for all along Western lines.

We should be ready to help countries living under near-Communist and outright Communist regimes; for we should realize that some such dispensation as this may be the inevitable price of the forced march that these countries have to make if they are to try to catch up with us. And we should also have faith enough in our own way of life to believe that, if we do give a helping hand to peoples who have been compelled by a temporary necessity to put themselves under non-Russian totalitarian regimes, they will take to our Western liberty just as soon as they find themselves able to afford it.

Part 3

THE SETTING
OF SOCIAL POLICY

TECHNOLOGISTS, I HAVE argued, are not truly in charge of much of anything, since their function is instrumental. They have been given considerable latitude, however, on the grounds that their labors produce clear benefits. It is the concern over costs, and, by implication, alternative courses of action, that has prompted urgent questions of social policy.

Policy is also plural, however, and politically determined courses of action are likely to command less than consensus. Democratic principles of distributive justice—Jeremy Bentham's "greatest happiness of the greatest number"—tend to be given at least lip service by most commentators on, as well as critics of, the course of technology, along with perpetrators of schemes that do not meet that test.

In the essays in this section, we encounter once more not just the counterpoint between the "natural" and the "social," but also between the technical and the humanistic. By deliberate design I have first put forward those authors least concerned with political policies and arrangements, and then those authors for whom such matters are a central concern.

We start with J. Bronowski, a scientist who is optimistic about the future of science and technology. He is mainly a professed believer in the mechanical and biological bases for a better future, but not naively so, for he gives "social revolution" full standing.

Brand Blanshard speaks eloquently for the humanistic tradition, preferring, if he has to make a choice, Hamlet to thermodynamics. Fortunately he does not have to choose, but for a reason that I fear he does not fully apprehend. The analysis of Hamlet or the cultivation of thought and reason in general could be enjoyed by very few were it not for the technologically produced affluence for which thermodynamics is his shorthand symbol.

Brooks Atkinson, a noted theater critic, apprehends the point clearly and develops it. Materialism, he observes, produces books and plays along with machines and industrial goods. The author essentially argues that we have too little cultural tradition to know that we cannot do the impossible, so we proceed to do it.

The historian Henry Steele Commager might almost qualify for the designation "social determinist." "The really interesting changes," he writes, "are not so much in material circumstances, or in the machinery of life, but in habits of thought and conduct, in manners and morals, in sentiment and taste." Commager's primacy of social changes does not help us much, for "they challenge and baffle our understanding."

The sociologist Herbert J. Gans, another social determinist, thinks he can at least contribute to our understanding. If I may rephrase and interpret his argument, it is that old ideals can be and are newly articulated and actualized, and that how affluence is produced becomes less interesting and problematical than how it is controlled and distributed. It is not at all clear that certain basic values have radically changed, but the physical and social means for their closer approximation in practice certainly have.

The political scientist Andrew Hacker plays some of the same themes, with variations. It is the *interplay* between democracy and technology that he finds responsible for our current state, good and bad together. His final sentence is a fitting close to the editor's tasks and to this book as such: "Our problems are those of success, and our failures are visible because we are continually conscious of the standards that we have set and failed to meet."

"1984" Could
Be a Good Year

by J. Bronowski

A WRITER WHO offers to forecast the future ought to begin by showing his credentials. My credentials are that I am an optimist and a scientist. I know that it is not usual for a prophet to be an optimist; most prophets prefer to play the part of Jeremiah and Cassandra. But then, that is because most prophets have not been scientists either; they have not really been in favor of progress.

We can see this in the most popular prophets of our own lifetime: in Aldous Huxley and George Orwell. Every reader must be struck by the revulsion against science and, joined with that, by the deep-seated fear of the future, which Huxley and Orwell share. "Brave New World" and "1984" are surely the most depressing societies that have ever been imagined, because their authors are so full of self-righteousness. They seem to me to be, not works of prophecy, but Puritan works of morality, preaching on every page a fire-and-brimstone sermon of foreboding. From the first page, the authors are sure that progress must be wrong—that everything that is good is already known to them.

I do not intend to follow the social and political preoccupations of Huxley and Orwell. Certainly the political world will be very different fifty years from now, when Asia and Africa will be immensely more developed and more vigorous in world affairs. But I shall not discuss politics, and I shall not even discuss social life in the future, except in one way—the way in which they will be shaped by the scientific discoveries and the inventions which can be foreseen now. I shall stick to predictions which are rooted in technical grounds.

There are three outstanding scientific changes which, I believe, will dominate the next fifty years. One is a change in the use of energy: this change has been set in motion by the discovery that men can tap the energy in the atomic nucleus. The second is a change in the control of energy: this change has been set in motion by the development of those electronic devices which go under the general name of automation. And the third is what I call the biological revolution: the discovery, which is still unfamiliar to us, that men can remake their biological environment, including parts of the human body and mind.

Nuclear Energy

One result of the addition of nuclear power to our other resources of power is, of course, to increase the amount of energy at the command of men the world over. What I have to say on this subject is best said in strictly numerical terms; and since I have already made these calculations once before, I should like to quote them as I made them:

"Today the inhabitants of the United States command, every man, woman and child, the amount of mechanical energy each year that would be generated, roughly, by ten tons of coal. This is the backing that civilization provides now, and it is equal, again at a rough estimate, to the work that would be done by a hundred slaves. If we are no longer a slave civilization, it is because even a child in the United States has as much work done for it as would require the muscle of a hundred slaves. By contrast, Athens at her richest provided for the average member of a citizen's family—man, woman or child—no more than five slaves.

"In most parts of the world, people today still command only a fraction of the American standard. In India, for example, the average use of energy amounts to the equivalent of about half a ton of coal a year, or five slaves—the standard of Athens over 2,000 years ago. This is the figure that will rise most steeply in the next fifty years. It cannot rise to the standard of the United States in that time, but it can reach a fifth of that standard. We can expect that in the next fifty years the energy used in the poorest countries will reach at least the equivalent of twenty slaves a head each year.

"It may seem very cold to measure the lives of people by the mechanical equivalent of two tons of coal a year, or twenty slaves. But the figure is not at all cold. In the first place, it could not be achieved had nuclear energy not been discovered. All the resources of the traditional fuels in the world would not yield this figure; and the dreams of liberal minds, to raise the dark races to the standard of the white, were an illusion until nuclear energy was discovered. The standards of the West will become at least tangible to the backward countries in the next fifty years because nuclear energy can provide the power."

This is one important effect of the coming of nuclear energy; and yet, to my mind, it is not the most important. To my mind, what is most important is that energy will be more evenly distributed in the future. It will no longer be necessary to concentrate industry where either coal or oil is plentiful. We shall not need to take the industry to the fuel, but the other way about— the fuel, the nuclear fuel, to the industry. For a nuclear fuel is more than a million times as concentrated as a chemical fuel; and where we could not take a million tons of coal or oil, to South America, to the copperbelt, to the Australian desert, we shall be able to take a ton of uranium or of heavy hydrogen.

True, it will, for example, still be proportionately cheaper to build a large nuclear power station than a small one. But there is now no longer an inherent difficulty in siting any power station far from the line of supply of its fuel. In the past, the logic of concentrating the generation of electric power in a few large stations was that it was easier to carry the current from the station in a wire than to carry the coal to the station in a truck. But if the fuel is nuclear fuel, this is no longer so; a small nuclear station can become the center of a remote township as effectively as it already drives a submarine or keeps an army camp alive under the polar ice.

Over the next fifty years, nuclear energy is also essential to the growing of food on a world scale. It is, of course, clear that if energy is cheap, then it is possible to make a substitute for any material that we need, all the way from industrial diamonds to vitamins. In this sense, then, we can count on finding a decent standard of living, in food as well as in energy, for all the six billion people who will be alive fifty years from now. And in agriculture, we shall need nuclear energy above all for the irrigation and exploitation of marginal lands, including the brackish lands now poisoned by salt water.

If energy is cheap and transportable anywhere, then irrigation is possible anywhere. In agriculture, what energy will buy fifty years from now is water; and water will be the key to growing food for the world's population, which will be twice as large as today.

Automation

Second, I want to discuss the future influence of automation. In one sense, an automatic machine is still a machine, and automation is no more than the logical use of machines. But, in fact, automation implies such a difference in outlook, such a change in the conception of the place of the machine itself, that I ought to discuss it quite fundamentally.

Two hundred years ago, the West discovered that a man's or a woman's output of work can be multiplied many times if the repetitive tasks which a hand worker must do are done by a machine. Machines were invented, all the way from the power loom to the mechanical digger, that could mimic those actions of the human muscle which a man must carry out laboriously and monotonously, time and time again, in order to get a piece of work finished. The wealth of the West, and its high standard of living, derive directly from the revolution in manufacture— the Industrial Revolution—which these machines created.

Until recently, the machines of industry confined themselves to those mechanical tasks which need muscle and no more. Only in the last years have we come to see, what now seems obvious enough, that any repetitive task is really best handled by a machine. This is true, whether the repetitive task is muscular, like rolling steel sheets, or whether it demands more delicate skills of calculation and judgment, such as controlling the thickness of the steel and computing its price.

This is the real nature of automation: the discovery that repetition is a machine task, even if the repetition is in adding up a ledger or controlling the distillation of a chemical. Men and women thrive on variety, but machines thrive on monotony. Machines do not get bored (and they seldom get tired), their attention does not wander, they do not feel that their gifts are being wasted. They like nothing better than to repeat themselves.

Fifty years from now, the machine operator of any kind will be as much a fossil as the hand-weaver has been since 1830. Today we still distinguish between skilled and unskilled jobs of repetition, between office worker and factory worker, between white collar and no collar. In fifty years from now, all repetitive jobs will be unskilled.

The social implications of this change are profound, and I believe that they, more than anything else that I have forecast, will shape the community of the future. For their effect will be to change the social status of the different jobs in the community. The ability to handle a column of figures will become no more desirable than the ability to drive a rivet; and even the ability to write business letters may become less sought after than the ability to repair the machine that writes them by rote. As a result, the clerk will sink in social status, and the electrical technician will rise; and that in itself is a change as far-reaching as was once brought about by the dissolution of the monasteries.

A Social Revolution

Here I want to turn boldly to make a social prophecy. I believe that the combined effect of nuclear energy and of automation will be to revolutionize the way in which men run their industries. Today industries are concentrated in large cities. The reasons for this are two-fold: we find it convenient to generate energy on a large scale, and at the same time we have to have large labor forces.

I have already shown that nuclear energy will make it possible to generate electric power in quite small units, where it is wanted. One reason for working in large cities will therefore disappear. But industry has moved to (and has created) large cities for another reason also: in the search, above all, for people. A product, whether it is a car or a can of polish, goes through many stages, so that many hands are needed to process and pass it on step by step. Is there any reason to think that industry will be able to break away from the huge arrays of semi-skilled workers which have served it hitherto?

I think that industry *is* breaking away, and that the traditional mass of factory hands *is* shrinking. The new wind in industry is automation, and I believe that it can transform industrial life in the next fifty years. There has been a great deal of technical talk about automation in recent years, but once again its more remote but important social consequences seem to me to have been missed. Yet automation is likely to revolutionize the balance between work and leisure, and the size and structure of community life, in the next fifty years.

Our industrial civilization has gone on herding people together in huge complexes of cities. Now there is a hope that the next fifty years may reverse this trend, and may begin to dissolve the ugly concentrations of the Ruhr and the Clyde, of Pittsburgh and Tokyo. In automation, joined to nuclear energy, we have the means to run industries on the scale of a small country town —a scale which does not dwarf the human sense of community.

This shift in the pattern of working life is the most far-reaching change that I foresee. The fifty years ahead of us will provide the means to create a social revolution: to create lively and efficient small communities which can hold their own in the industrial world.

There are many things to be gained by leaving the large cities. For example, we shall gain the hours (about one eighth of our waking hours) which most workers now spend in the tedium and discomfort of travel. This is a great gain in leisure, and some people will think that it will create new problems of leisure. I do not think so; I think that leisure is only a problem today in those places where tedium and discomfort have reduced everyone to a dull indifference.

I am not the first prophet, or the first dreamer, to hope that the monstrous cities of today, like glaciers of an industrial ice age, will begin to melt away. But when social reformers in the past have longed for small communities which could be self-sufficient, they have usually wanted to found them on agriculture. They have wanted to go back to the land literally—to work on the land. This is quite unrealistic, now and in the future.

In short, it is not necessary to retreat from the disaster of the metropolis into the inertia of the village. The small town of the future can be as well-equipped, physically and intellectually, as the largest modern city. It will be served physically by the new forms of travel, and intellectually by the new links of communication which we can already foresee. My guess is that it will then need to be large enough only to support those unpractical but delightful luxuries which give life to a community—a baseball club and a theatre and places where people play chess or go bowling. I think that you can do all these things in a town of about thirty thousand strong, and this is my forecast of the size of the new industrial communities in the future.

But the small community that I have sketched has no room for dullness and indifference. If thirty thousand people make an industrial town which is physically and intellectually self-sufficient, they must all be skilled. The one unpractical luxury that they cannot afford is a man with no skill.

The Biological Revolution

The third fundamental change which will, I think, shape the future is what I have called the biological revolution. We are just beginning to learn that we can mold our biological environment as well as our physical one. During the next fifty years, this will be the most exciting and, I believe, the most influential work in science.

Let me single out a few lines of work which seem to me especially interesting and promising.

There is, to begin with, the practical progress in the attack on organisms which damage us. They may be pests which damage our food supply, at one extreme, or microbes which invade our bodies at the other. Fine work has already been done in developing specific chemicals to tackle each specific enemy. I think this method of combat, the development of exact and specific chemicals, will play a growing part in making men healthier and richer.

Let me give one example. We used to think that a man could produce the antibodies which resist a virus infection (for example, smallpox) only if he were given a mild dose of the infection. Now we know that this is not necessary. We know that a virus consists of two parts—a living center of nucleic acid, and an outside covering of protein. And we know that the protein covering alone will suffice to stimulate the cells to produce the antibodies which fight the whole virus. This is how, for example, the protein in the killed polio vaccine works.

I believe in the future we shall go even further: we shall protect against a virus disease by making, in the chemical laboratory, the protein covering of that virus.

This leads me to the next outstanding field of study. We know that there are drugs which greatly sharpen a man's faculties, and others which help him to be at peace with himself. Each kind of drug helps a man to make better use of his natural (but often hidden) gifts. I am sure that there is a bright prospect here for the future and that, as a result, men and women will lead livelier and happier lives, in work and in leisure right into their old age.

Finally, I should pay tribute to the searching work that is being done in the study of biological processes on the smallest, molecular scale. This has already given us a new understanding of the nature and of the dynamics of life, and at this very moment it has opened a deep insight into the basis of all heredity. I believe that in the long run this fundamental knowledge, which still seems abstract and remote, will have the greatest effect of all in the practice of medicine.

Science for Peace

I began by saying that I am an optimist and a scientist, and you now see that the two go together. There is plenty of ground for pessimism in world affairs, and perhaps we shall not avoid the suicide of mankind. But can we not? Can we not prevent the leaders of nations from being proudest of those scientific inventions which make the loudest bang?

We *must,* exactly because science has so much better uses to offer for its fundamental discoveries. I have shown you the rich future that should grow out of the very discoveries that people dread most—out of nuclear energy, automation, and biological advance.

What people fear is the reach, the power of these discoveries. And there people are not foolish: they recognize that nuclear energy, automation, and biological advance are the most powerful social forces of this century. But that power can be as great in peace as in war; we can use it to create the future and not to murder it. Science promises a future in which men can lead intelligent and healthy lives in cities of a human size, and I think it is a future truly worth living for.

"Hamlet" vs. the Laws of Thermodynamics

by Brand Blanshard

WE HAVE BEEN warned of late that we need more science and far more scientists. Granted. But perhaps it is not untimely to recall that there are such things as humanities also, and that we shall do well to keep them, too, alive.

People throughout the country have been reading Sir Charles Snow's little book "The Two Cultures," and in many of our cities and universities the author has been making personal appearances. With the special knowledge of a scientist and the skill of a man of letters, he builds a powerful case for the sciences. He says—what is obvious enough—that if we are to hold our own in a military way against the Communists, we must have more and better men of science. He contends—what is less obvious but also true—that if we are to prevail with the vast, hungry, restless, ambitious, uncommitted peoples, we must be ready to send them regiments of scientifically trained helpers.

That means that we must multiply our scientists. And if we are to do that, we must first become aware of our scientific illiteracy and then we must make a general effort to remove it.

From the *New York Times Magazine,* December 24, 1961, copyright © 1961 by The New York Times Company.

Are we scientifically illiterate? Too nearly so, beyond a doubt. Sir Charles reports that he once asked a company of cultivated persons how many of them could describe the second law of thermodynamics, and that their response was both negative and chilly. Yet this, he says, "is about the scientific equivalent of 'Have you read a work of Shakespeare?' " He goes on to say: "I now believe that if I had asked an even simpler question such as 'What do you mean by mass, or acceleration?'—which is the scientific equivalent of 'Can you read?'—not more than one in ten of the highly educated would have felt that I was speaking the same language." Finally, he suggests that even at a Cambridge high table, if someone had mentioned Yang's and Lee's experiment at Columbia, whose result was the contradiction of parity, the "two cultures" would have looked at each other across the table with blank faces.

All this, I agree, is deplorable. We who count ourselves more or less educated ought to know the elements of science, and many or most of us do not. And it is vastly important that we should have Yangs and Lees who know about the principle of parity, and the evidence for and against it.

But we need to keep two questions distinct. The first is: what kind of education is most useful in a hot or cold war? To that question the answer may well be science. The second is: what kind of education does most for the individual mind? To that the answer, I think, is the humanities.

And it is a tragic delusion to suppose that in answering the first question, we have also answered the second. If you ask me which I would rather have about, for his own sake and the community's—a man who knew what mass and acceleration were and even the second law of thermodynamics and the principle of parity, or a man who could enter fully into "Hamlet" and "Lear" —I would choose the latter.

The pursuit of understanding is fundamental, I agree, and science helps in that pursuit by providing us a discipline in straight thinking. But the difficulties of thinking vary with its subject-matter, and the question of the liberal-arts student is: does science equip us against the particular kinds of difficulty that most of us will meet in our thinking when the college gates have clanged behind us?

Well, what *is* the sort of thinking we shall have to do? It is the thinking necessary to the good citizen, the good neighbor, the good father or mother of a family, the competent man of affairs, the supporter of sound causes generally, the person with sensitive allergies for political hocum, specious advertising, religious superstition, class and race tension, and lopsided partisanship in all its fifty-seven varieties.

In this kind of thinking the difficulties are less logical than psychological. The main one is to stick to the path of reasonableness through a fog of passion, pride and prejudice, to thread one's way through a thicket of likes and dislikes, callow enthusiasms, dubious authorities and distorting complexes.

Now the thinking of physical science is not this kind of thinking, nor does practice in its clear air necessarily help us much in the murky ways of common life. The physical laboratory is disinfected of emotion; among its equations are no personal equations; its atmosphere is air-conditioned to exclude the mist and miasma that surround us in the house and on the street. The scientist seeking to verify or amend Ohm's law may have to struggle against his own slowness and thickness of wit, but he does not noticeably have to struggle with superstition. Ohm's law is either true or false; emotionally it does not matter which; why be hot about it?

The case is quite different in the humanities. Of course, there are men claiming to be humanists who, like Browning's grammarian, spend years on "the doctrine of the enclitic *de,*" but this I decline to accept as humanistic study at all.

I am thinking, for example, of history, where, if we are to understand the French Revolution or the Civil War or the rise of Hitler, we must control our impulsive partisanship; of political theory, where we cannot begin to understand the hold of Karl Marx over the minds of millions without an effort at enlarged sympathy; of literature and art, where an appraisal of "The Wasteland" or "Ulysses" or Ginsberg's "Howl" may force us to examine our hierarchy of values; of philosophy, where, if we are to think straight about God or freedom or immortality, we must fight our way through a thicket of prejudices as old almost as the race.

The best training in objectivity lies in fields where there is some temptation to subjectivity, just as the best training in clearness of thought lies in those fields where clearness has to be achieved by you and is not waiting for you ready-made. That is the difficulty with mathematics as an educational discipline. I admire at a great distance the skill of the mathematician in manipulating his symbols according to his recondite rules, just as I admire the astonishing gift of young Bobby Fischer for manipulating the men on a chess board. But I confess to some disquiet when I learn how much of this sort of "thinking" can be done, and done more surely and swiftly, by machines.

The greatest living mathematician, Lord Russell, reports somewhere that the thinking of "Principia Mathematica" was largely done with his fingers and that, as for thinking of the more concrete kind, a mathematician could get on with a few minutes a month. How these matters really stand I am not qualified to say. But even a tyro knows that when you are dealing with numbers or triangles, you are dealing with concepts that are perfectly definite already.

You do not have to hew your way to clarity at the outset, as you do when you try to think about liberty or democracy, beauty or goodness, conservatism, Catholicism or communism. Yet these are the concepts, with all their mistiness and emotional charge, which we must manage to deal with as men and as citizens.

Perhaps more important: science is exclusively intellectual; liberal education neither is nor ought to be. Granting that the impulse to know is for education the central one, there are twisted and festooned around it a mass of other impulses also essential to our humanity, upon whose fulfillment the richness of life depends.

Besides being a thinker, man is a doer and an appreciator. I doubt whether a college can give him much help as a doer. Even as a thinker that is hard enough: "You can lead a girl to Vassar, but you can't make her think." It is harder still on the side of action, for there is no adequate classroom test for character and firmness of will.

But as appreciators, it is clear that students can be emancipated and sensitized. Just as their discernment can be sharpened for the true and the false, so their feelings can be made more discriminating as between the good, on the one hand, and the cheap or tawdry or ugly, on the other. Culture has been defined as the right adjustment of feeling to its objects.

It may be held that in such fields as music and painting, poetry and fiction, good and bad are matters of taste, that about taste there can be no disputing, and that talk of a really better or worse is therefore dogmatism. To the many who nowadays say this sort of thing, one can only reply: Why not go on and draw the logical conclusion? Why not pitch all these subjects out of the curriculum? What is the point of spending time and money in forming tastes that are really no better, by your own admission, than those the students brought with them in the first place?

But, of course, we all have standards and use them, even when we disavow them in words. For a given man there will be that which is really good, and it will be what, with due exposure to it, will yield him the completest fulfillment and satisfaction of which, with his nature, he is capable.

We never know in advance and with certainty what that is. But we are not merely in the dark about it either. For some seventy generations people of the West have been reading books, looking at pictures and listening to their music-makers. In each generation they have found a few who spoke to their condition. These few they have treasured and passed on to their children. The children have compared them with those of their own generation who did the most for them, and handed on the increased anthology to their children in turn.

By now we have a list of true classics in Dr. Hutchins' sense of "works contemporary with every time." If someone says that he finds these classics nauseating, as Spencer did of Homer, and Darwin of Shakespeare, we feel that the remark says much about the speaker, but not much about Homer or Shakespeare.

The aim of the humanities is not to improve our powers of analysis and argument only, though how an understanding reading of Plato or Mill could fail to do this is hard to see, but also to make us whole men. And a man is not a whole man, he is a maimed and stunted man, if he is blind to Rodin, deaf to Mozart, indifferent to what Eliot, Gropius and Picasso have been trying to do, bored by drama, unseeing of the white trees in spring and the scarlet ones in autumn, unmoved by Schweitzer, as stolid in a cathedral as in a barn.

The aim of the humanities is, if the phrase is not too tired by this time, the more abundant life. This aim they try to achieve by the simple process of exposing us under auspicious conditions to what responsive people have in fact found to open a larger life to them. The humanist cannot prove to the skeptic in advance the value of his wares. He can only say that many who have tried them have been able to report, as he has himself, "Whereas I was blind, now I see."

Of course, the humanities often fail. Sometimes they do worse than fail; badly taught, they may arouse a permanent antipathy. But the sciences do not even try to awaken this particular kind of vision, and quite rightly. Kepler did, to be sure, exclaim that he was thinking God's thoughts after Him, and the early Russell sang paeans to mathematics (which the older Russell only faintly echoes); but it is a dangerous business, in Santayana's phrase, to "fuse one's physics with one's visions."

Science leaves values alone, except the value of truth, but men cannot afford to leave the other values alone, and therefore a scientific education is a defective one.

The great educational need in this country is not for more science but for more of what the humanities are at least trying to give. Of what science has to give we have more than any other country in the world, and from the material excess that science has made possible we are distributing billions in largesse to needy neighbors.

To our astonishment and chagrin those neighbors seem to want it only on condition that we do not seek to send with it our American ideals and way of life. These, thank you, they do not want. They have come to think us in all essentials as materialistic as the Russians, a drivingly efficient and successful people, but crude, noisy, and immature.

It is an unjust and an ungrateful image of American life, the product largely of envy and rationalization. But there is enough truth in it to hurt. The picture of the overfed American, sitting in his air-conditioned living room and drinking his Coke before his expensive television set, his mind intent on an adolescent Western, or whiling away a boring Sunday morning with the comics and sports section, is the symbol of a strange combination of technical achievement and cultural vacuity.

Our movies, our best-sellers, our recordings, our adulated Elvis Presleys and Marlon Brandos, the glossy books and magazines that fill our station newsstands, do not give a very flattering testimony to popular values. And this in spite of the fact that, if expenditure and numbers count, we are the best-educated country in the world.

I take it that T. S. Eliot, in view of all the Sweeneys and Prufrocks, would abandon the American dream and go back if he could to a hereditary aristocracy and a religion on the Anglo-Catholic model. If this were the only salvation from universal vulgarity, we should no doubt take it seriously. But some of us have not lost hope of what W. C. Brownell called "democratic distinction in America."

In our high schools and colleges we have a splendid set of channels for disseminating values; what we now need is men and women staffing them who are in a deep sense humanized themselves. Is it visionary to hope for the day when a Frenchman or a Burmese, hearing that Jones is an American, will automatically infer, not "then he will be well-dressed, rich, and hard," but "then he will be kindly, just and understanding?" There is no harm in dreaming dreams.

Please do not mistake me; to say that we need more of the humanities is not to say that we need less of science; clearly our colleges must retain and develop both. If Sir Charles Snow lays so heavy a stress on science, it may be in part because the student who pursues humanities at Oxford and Cambridge gets no instruction in science at all. This is not true at our own universities; all students are expected to have some acquaintance with science, though most pursue some other subject as their major interest.

This, I think, is as it should be. My point is that it would be a blunder, and in America a peculiarly unfortunate blunder, if the military primacy of science should persuade us to put it first as a tool of education. Whatever else it may be, *this* it is not.

Where Sir Charles stands on this strictly educational issue I do not know. But we may know soon. He has just been elected Lord Rector of the University of St. Andrews, and on this issue one suspects that he will agree with an illustrious predecessor in this office. About a century ago John Stuart Mill gave the rectorial address at St. Andrews, and he chose as his theme this very issue: which is more essential in education, the sciences or the humanities? His answer, after fifty pages of meticulous discussion: both.

In Defense of
American "Materialism"

by Brooks Atkinson

ALTHOUGH AMERICAN culture is not highly esteemed in many parts
of the world, American plays and books are only a little less
popular than American goods. Particularly in this post-war era,
many of our novels, dramas and books of thought are hospitably
received abroad; and, of course, there is an abnormal demand
for our industrial goods.

I am deliberately associating our literary products with our
industrial goods: they are integral parts of the same culture.
What a man does is as much a part of his culture as what he
thinks. A culture is the sum of all of the activities of all the
people who compose a nation; and in democratic society there is
no arbitrary difference in rank between the doers and the thinkers. Character is a stronger influence in a society than intellect,
and character is a common element in all spheres of life.

We all live together and eat from the same granary (which is
probably the fundamental source of a national culture); and I
do not believe that the scholar or littérateur categorically out-
ranks the farmer and the mechanic. Quite the contrary: the

From the *New York Times Magazine,* April 16, 1950, copyright © 1950
by The New York Times Company.

scholar and the littérateur lose their usefulness to society and become nothing more than barren symbols if they are arbitrarily dissociated from the farmers and mechanics.

Americans have always resented and resisted authority. The tradition of scholarship as a separate and rather exalted department of life has never taken a firm hold on American culture. In the early days of American history, the demand was not for members of the genteel classes but for farmers and artisans—for carpenters, masons, blacksmiths and husbandmen.

Obviously that condition rose directly from the urgent needs of a primitive colonial territory that was being developed in a hurry and instinctively was taking its art, literature and scholarship from abroad. But the separation of the thinkers and doers never has been accepted in America. "Life is our dictionary," Emerson said in 1837 in "The American Scholar." "There is virtue yet in the hoe and the spade for the learned as well as unlearned hands." For Emerson, a democratic idealist, realized that the life of man and nature is the primary source of thinking, dreaming and creating; and that art and scholarship are bogus unless they derive from life directly.

Contact with life is the first principle of thinking and creating. Contact with libraries, museums, lecture halls and theatres is hardly more than the secondary stage. For unless art and scholarship are constantly renewed in the streets and fields, the basic errors in thought and the idle whims of the imagination can never be discovered or corrected, and theories, books and plays will have treacherous foundations. The separation of the thinker from the doer weakens the whole structure of society.

During the war in China I remember being amazed and baffled by the social divisions of a group of young Chinese officers who were studying veterinarianism. They wore the long white coats of the scientist and carried fountain pens, notebooks and textbooks. But they seemed to be squeamish about touching the animals. They left the animals to the private soldiers and the non-commissioned officers. Social customs that are probably thousands of years old inhibited them from working directly with the raw material they were studying.

I do not know whether this was an isolated instance or common practice among student veterinarians. But it followed China's ancient systematization of the classes, which venerates the scholar, gives him unusual social prerogatives, relieves him of many of the routine chores and associations of daily life and thus saps his personal independence. In the painful process of modernization, China will doubtless weaken or destroy the old departmentalization of social life and release the energies of educated people for broader and deeper living.

No culture is vital today if it has the effect of leaving large areas of the population in ignorance or servility or malnourished. Old wisdom, like that of the ancient Chinese sages, supplies a cultured defense for corruption, lethargy and the bondage of the people. You can always find a picturesque old saw in any culture to excuse or defend intolerable conditions. Stuff a man's mind with enough old sayings and he will never wrestle with a new idea. Thousands of young Chinese, as well as many of their elders, already know this and are acting on it as the modern premise.

In America we lack almost entirely the classical tradition. But from the pragmatic point of view I wonder whether this may not be a good thing and whether it does not foster the flexibility and optimism of American life. We are not resigned to repeating in the future the tiresome disasters of the past.

Let's not underestimate the material parts of a national culture. To look down on them disdainfully is to eliminate from good society the vast bulk of the population. Without a sound materialistic underpinning the rest of the culture is hardly more than an empty pretense or a whimsical fairy story. People have to be fed, which is the province of the farmer. People have to be housed, which is the province of the building mechanics. Since people also like to be fastidious about their personal habits, keep warm in the winter, keep their food fresh in the summer and move as quickly and easily as possible from one place to another, the plumber and the manufacturers of heating plants, refrigerators and automobiles perform useful functions in modern society. Taken together with the millions of trained people they employ and also the farmers, they constitute the largest group of people in the country.

Since the process of manufacture includes many thousands of minute mechanical operations, industry has a low esthetic rating. The whole of industry is sometimes thoughtlessly dismissed as a kind of odious and inhuman routine, and the American civilization is ticked off as "machine-made," as though it were mediocre or soulless.

But there is a flaw in that thinking, a confusion of the process with the creative idea. Behind every machine stands a man who imagined it for a particular use in modern life. The basic idea may have been as valuable and enriching as the basic idea of a book or play—indeed, it may have been of greater value, since the ideas in many books and plays are poverty-stricken.

As a product of human creative energy, the high-speed newspaper printing press is particularly worth respecting. It is daring in conception, ingenious in design and perfect for what it is intended to do, which is to give fast and wide expression to information and opinion. Although it is made of steel it is a thoroughly human enterprise. A human being imagined it; a great many human beings have made the drawings and moldings to build it. Human beings operate it and keep it in repair. There is nothing soulless about it: it is charged with human energy and spirit. Deprive it of human beings for an hour and see how long it could work. It is a tool that expresses part of the human spirit, like the piano, brush, crayon and scalpel.

In its useful forms, materialism is a product of spirit. The drive of a culture impregnates it and radiates from it. In wartime the central idea of a national culture is expressed through it directly. In their pre-war analysis of America's fighting ability, the Japanese believed that Japanese spiritual tenacity would win because the Americans were too materialistic for the test of battle.

But the materialistic abundance and ingenuity of the American fighting forces were in themselves the products of spirit—of the free spirit, in fact, which reached out in all directions. Ships were built at high speed in great numbers. Goods of all kinds were manufactured in staggering quantities and delivered thousands of miles away. The supply service was so perfected that once on Okinawa a few lucky marines under fire in advanced positions had hot meals of steak and French fried potatoes.

All these things and millions of others did not represent the mechanical versatility of robots. They were direct expressions of the American spirit brought to bear on a particular problem under emergency conditions that aroused the vitality of all the people. Behind the deadly fire power that swept the Pacific was the tenacious spirit of free men who would not submit to tyranny. Materialism is decadent and degenerate only if the spirit of the nation has withered. The machines that are now in existence and working efficiently are part of the capital wealth of the nation. But the imagination, skill, ingenuity and energy that have gone into them are part of the national culture.

Since America has instinctively resisted the dissociation of art and scholarship from the elementary functions of society, I think our plays and books radiate the energy and imagination of the people. They are part of the mass culture. There is no such thing as "closet drama," which is too refined for the general public. Cotérie art has no real importance or significance. Scholars are not venerated as a social class, but are under the same compulsions as other people to demonstrate their value as good citizens and neighbors by their understanding of human beings.

Officers of the Government, including the President, have no prestige beyond what they earn for themselves by intelligent leadership and good conduct in office. In fact, the criticism of public officers is often so bitter and malevolent that sensitive and high-minded people hesitate to leave the protection of private life and expose themselves to the savagery of political opponents. Although the disciplines of democratic culture are basically wholesome, they are frequently painful and often unfair.

It is significant of our culture that our only indigenous philosophy is the pragmatism of William James and John Dewey, which evaluates ideas according to their usefulness. Pragmatism suits the American philosophy, and is probably the working philosophy of many of those Americans who hunger after idealism in their thinking. In the final analysis, Americans are not much impressed with theories. Economically the country has outgrown its youth, but its culture represents a logical development from the independence, enterprise and willingness to work that characterized the pioneers who settled here two or three centuries ago.

American literature is only a little more than a century and a half old, and perhaps not quite that; and American drama is less than a half century old. By and large, American art reflects the mass culture of the country, which is still new. Nations with a culture developed out of the classical tradition are often troubled or confused by the impulsiveness and often the violence of the culture that comes from America.

But they are also inclined to like not only the machines and industrial goods but the books and plays. These are different aspects of the same thing. Everything American, whether it is thinking or doing, comes out of the same reservoir of energy. And I hope that the normal goodwill of the American character compensates for the lack of decorum in our culture. For the truth is that Americans do not value the proprieties highly.

We Have Changed— and Must

by Henry Steele Commager

WE ARE IN the midst of changes at home and abroad, as far-reaching in their implications, and in their demands upon us, as any we have experienced in the past. Some of these flow from scientific and technological developments connected with the harnessing of atomic energy, desalinization and the redistribution of water, automation, and so forth; some are the consequences of the new world pattern created by the emergence of Africa and Asia into modernity. Some are created by—or perhaps just required by—the growth of population, cities and the welfare state.

This is an old story, yet one that is ever new. How extraordinary, we assure each other, the changes that we have experienced, the changes that our society has experienced: from the horse and buggy to the automobile; from the railroad to the jet airplane; from dependence on the vagaries of weather to reliance on oil heating and air-conditioning; from stage shows to television; from old-fashioned guns to atomic missiles. The list is inexhaustible.

A good many of us never get much further in our contemplation of these changes than an expression of astonishment and—

From the *New York Times Magazine,* April 30, 1961, copyright © 1961 by The New York Times Company.

usually, though not always—of gratificåtion that it should be our fortune to live in a civilization so mechanized and improved.

Yet these changes, far-reaching as they doubtless are, and fascinating, too, are almost purely quantitative. Certainly since the industrial revolution each generation has experienced them—and almost always with comparable excitement. It is a pretty safe prediction that each future generation will continue to experience them. Let us not disparage them. They add immensely to the interest of life; they maintain the economy; they excite the imagination; they give an illusion of progress and, in some instances—medicine, for example—the reality. But they are not, in fact, very surprising. If they did *not* materialize—that would be really surprising. Nor are they, in any intellectual sense, deeply interesting, for they are the commonplace of growth and of time.

There is a second category of change that is more interesting and more significant, but, again, not particularly puzzling. I refer to those changes, often basic and far-reaching, which are brought about by shifts in the economy or in the political machinery. These changes are for the most part self-explanatory.

Thus, for example, the passing of thrift: a product of inflation and of the welfare state. Thus the changing position of the military: a product of two World Wars, of dependence on the military for survival, and of the shift in the center of gravity of the military itself from the battlefield to the factory and the laboratory. Thus the passing of the myth that rural life was somehow morally superior to urban life: with the population three-fourths urban or suburban, and the countryside itself largely urbanized, and the advantages of urban life so plain, it is inevitable that philosophy should adjust itself to fact.

These and similar changes in the mechanics of life do not present any serious challenge to the understanding.

The really interesting changes are not so much in material circumstances, or in the machinery of life, but in habits of thought and conduct, in manners and morals, in sentiment and taste. In the long run these may well be the most important, as well, for they dictate, or at least condition, changes in the economic and material arrangements of society. What is more, they are irresistibly interesting, for they do not lend themselves to easy explanation, but challenge and baffle our understanding.

Let us look at a few examples of these changes.

First, consider the growth of humanitarianism. Our great-grandparents—say, a century ago—were at least as virtuous, as religious, as kindly and humane, as we are. Yet they tolerated, nay, took for granted, what we would consider monstrous in-humanity of man to man—and to woman and to child, too. They took for granted that those who were unable to pay their debts should languish in prison for months and sometimes for years. They condemned the feeble-minded to imprisonment in wretched cells, chained them to walls, beat them for their failings and starved them, too: the whole story can be read in Dorothea Dix's famous report of 1842.

They allowed little children to work twelve hours a day in factories and mills. They inflicted brutal punishments on prisoners and seamen—flogging, for example— and they assumed, too, that teachers would keep order in schoolrooms by the liberal use of the rod. They condemned immigrants to life in miserable hovels that were breeding places for vice and crime and disease. South of the Mason-Dixon line, Christian men and women not only tolerated the enslavement of Negroes, but counted the "peculiar institution" a positive blessing for all concerned.

All this has changed. Public opinion no longer permits children to work in factories, no longer tolerates mistreatment of prisoners and the feeble-minded, and has done away with flogging in the Navy and the schoolroom. And even the most intransigent of White Citizens Councils would be appalled at the suggestion of reviving Negro slavery.

Or consider the change in the position of women. A century ago it was taken for granted that while women were morally superior to men they were intellectually—and in almost all other ways—inferior. Like children, they were to be seen but not heard. The London meeting of the World Anti-Slavery Convention broke up because the British would not permit lady aboli-tionists to participate as delegates!

Women were allowed to work long hours in factories, but not

to practice law or medicine, or to preach. The Army did not even want them as nurses in the Civil War. Married women had, in effect, no rights that their husbands need respect: no right even to their children. Everywhere the double standard of morality was taken for granted. As for politics, as late as 1912 intelligent men were gravely prophesying the disintegration of society and the collapse of morality if women were so much as allowed to vote!

In less than a century all this changed. The double standard gave way to the single. Women today not only control their own property, but the major part of the wealth of the nation. They are not allowed to work long hours, or at night, or for less than minimum wages, but there is no profession from which they are barred, and few which they do not adorn. And none of the dire consequences of their participation in politics has materialized.

Or look to a quite different, but no less significant, development: the growth of tolerance in the past century or two—a short period, after all, in human history. From time to time we are alarmed, and justly, by manifestations of intolerance in our society—by anti-Semitism or anti-Catholicism, by racism in the South, by McCarthyism, by dangerous pressures for intellectual and social conformity. But if we look back over a period of a few centuries what is most impressive is the steady growth of tolerance—in religion, above all, but in politics and society as well.

Time was when every nation had a State Church and, what is more, enforced conformity to it by rope and fire. Nor did governments tolerate dissent in the political realm. In the seventeenth and eighteenth centuries, and even in the nineteenth in some countries, political dissent was silenced as fiercely as religious.

Criticism of the King, or of the State, was commonly treated as seditious libel. As late as 1663 William Twyn was drawn and quartered for "imagining" the death of the King, and a few years later Judge Jeffreys sent the noble Algernon Sidney to the gallows for writing an unpublished manuscript advocating republicanism. Persecution persisted through the eighteenth century and into the nineteenth in England: as late as 1850 Catholics could not attend, nor nonconformists teach, at Oxford or Cambridge.

Nor were our own forefathers much more tolerant. The Puritans of Massachusetts Bay drove out Roger Williams and Anne Hutchinson, and persecuted Quakers and "witches" as did the rulers of Virginia. During the Revolution patriots decreed death for Loyalists, and conveniently confiscated their property and not long after the Revolution Congress made it a penal offense to write or publish anything that was designed to bring the President or the Congress "into contempt or disrepute."

A generation or so later Southerners zealous to end all discussion of Negro slavery banned magazines and newspapers, burned books, purged libraries, silenced teachers and preachers who agitated the slavery issue. In 1834 a mob in Charlestown, Mass., burned an Ursuline convent; a jury acquitted the mob leaders and the state refused to compensate the Catholic church for the destruction of its property.

In 1837 a mob in Alton, Ill., killed the abolitionist Elijah Lovejoy and destroyed his press. The following year a Philadelphia mob burned Pennsylvania Hall to the ground because abolitionists held meetings there.

Clearly the climate of opinion has changed. There is no active religious intolerance in England or in the United States now, no suppression of political discussion or even of social nonconformity. Who would have thought, at a time hardly more than a century ago when mobs were burning Catholic convents, that it would ever be possible to elect a Catholic to the Presidency?

The most baffling of all changes are changes in taste—whatever that elusive word may mean. An infallible way to induce gaiety and a comfortable sense of superiority is to show pictures of bathing costumes of the Eighteen Nineties, or of domestic architecture and interior decoration in the days when the stained-glass window, the rubber plant and the antimacassar were the epitome of good taste.

How does it happen that our grandparents took delight in architecture with turrets and towers and spires and fretwork and stained-glass; that our grandmothers filled their parlors with potted palms, hung heavily tasseled velvet draperies between rooms, and put paper fans in the highly decorated black marble fireplaces? Why did they turn away from the simple and dignified architecture of the Federalist period or of the Greek Revival toward the pseudo-Gothic, the pseudo-Italian and the pseudo-Queen Anne?

Will our grandchildren be as repelled by—or amused by—our glass office buildings, our ranch houses, our functional furniture and our decorative austerity? Beauty, we know, is in the eye of the beholder; why is it that the eyes of each generation reflect such different visions of beauty?

Look, finally, to a major social change—the change in the attitude toward work and toward play. For three hundred years —that is, almost up to our own time—Americans took work for granted and, what is more, looked upon it as a blessing.

Benjamin Franklin's Poor Richard spoke not only to his own generation but to future generations in his many admonitions to be up and doing: "Diligence is the mother of good-luck"; "God gives all things to industry, then plough deep while sluggards sleep"; "The used key is always bright"; "Industry pays debts"; "The sleeping fox catches no poultry"; "There will be sleeping enough in the grave"—these and dozens of others were household axioms for two centuries.

Not only in Puritan New England and Quaker Pennsylvania, but everywhere in the country, in the South and along the frontier as well, it was assumed that work was the destiny of man and that it was on the whole a good destiny. Few Americans of the past were disturbed by long hours of labor; few gave thought to vacations; and the cliché "Relax!" had not yet entered the language. Play was well enough for children—though even children were admonished that Satan found mischief for idle hands—but when boys and girls had passed the age of 12 they were expected to work like their elders.

Only in the last generation has work come to seem the exception rather than the rule, something to be avoided rather than something to be embraced. The ideal of our time is "relaxation." The coffee break and the cocktail hour, the long summer vacation, the ski week-end, the winter trip to the Caribbean; the cult of sports and of games, the evenings devoted to bowling or to television; the rise of the country club to the position of a national institution—all these reflect the American mania for play.

Nor is this merely a response to automation or to labor-saving devices. The rich of earlier generations had their labor-saving devices as well: cheap labor and slave labor; but that did not persuade them to try to make their lives a perpetual vacation, or free them from a sense of responsibility to work to the limit of their capacity.

How account for these, and similar, changes in moral habits and manners and taste? We cannot explain them but we can, in a sense, bound them, as we used to bound states or countries in old-fashioned geography.

Change is a phenomenon of a highly civilized society. The American Indian, the Aztec and the Inca did not change, nor did the Hindu or the Bedouin during the past three or four centuries. It is a phenomenon of the modern world: there was, apparently, very little change, or interest in change, before the Renaissance.

Change appears to be associated with three major modern institutions—the city, the machine, and democracy. Certainly it proceeds more rapidly and more easily in industrialized economies than in agricultural. It is associated with cities—the very climate of city life encourages experimentation, as well as indifference to tradition and to the past. It is related to education because education presumably opens minds to new ideas; and to democracy because democracy enlists the average man in the affairs of his society and thereby encourages discussion.

These characteristics—industrialization, urbanization, education, democracy—are peculiarly prominent in American society. Historically, Americans have been the people most tolerant of, indeed enthusiastic about, change. That process of uprooting and transplanting which was the settlement of America is the obvious and dramatic manifestation of this. But it is in the realm of society, economy and politics that Americans proved themselves most resourceful, most ready to challenge the traditional and embark upon new enterprises.

They challenged the notion that men were to be governed by kings and nobles and priests, and set up the first large-scale experiment in self-government. They challenged the principle that church and state were two sides of the same shield and that each was essential to the support of the other, and set up a state without a state church.

They challenged the age-old notion that society was divided into classes whose position was part of the cosmic system— "untune that string, and hark what chaos follows"—and they inaugurated an experiment in a classless society. They challenged the notion that colonies were always subordinate to a mother country and designed for exploitation, and were the first people to do away with colonialism altogether.

There is no such thing as standing still; to cling to the past, or try to preserve the status quo intact, is to go backward. Change does not necessarily assure progress, but progress implacably requires change. If our society is to flourish and prosper, it should encourage both institutions and practices that facilitate change: above all, the growth of education and of free discussion.

Education is essential to change, for education creates both new wants and the ability to satisfy them. It inspires at once that discontent for existing conditions and that faith in improvement which are essential to progress; and it provides the technical skills that enable us to achieve the goals we set ourselves.

Also, if we are to progress by evolution rather than by revolution, we need to encourage free discussion with no quantitative or qualitative limits. Those who fear change instinctively try to suppress the give-and-take of free discussion. They delude themselves that they can achieve security and maintain things as they are by smothering curiosity, blocking inquiry and silencing criticism.

But that policy has never assured either peace or security, unless it is the peace of stagnation or the security of death, and inevitably it drives discontent underground. We have experience

207 • *We Have Changed—and Must*

of this in our own history. Thus, leaders of the Old South deluded themselves that they could somehow maintain slavery by preventing any discussion of it, thereby imposing on the people of the South a uniform pattern of thought and conduct.

Of course, all they did was to drive discussion not so much underground as North, and to force those who hoped to ameliorate the great evil of slavery into drastic and violent measures. They did not save the institution—perhaps they could not—but they made sure that its liquidation would come about through violence; they did not save their society, but brought about its destruction.

There is no assurance that education and freedom will in fact enable us to solve those tremendous problems that loom upon us or assure a peaceful and prosperous future to mankind. But this is certain: that without education and freedom it will be impossible to solve these or any problems.

The "Equality" Revolution

by Herbert J. Gans

SOMEDAY, WHEN HISTORIANS write about the nineteen-sixties, they may describe them as the years in which America rediscovered the poverty still in its midst and in which social protest, ranging from demonstrations to violent uprisings, reappeared on the American scene. But the historians may also note a curious fact, that the social protest of the sixties has very little to do with poverty. Most of the demonstrators and marchers who followed Martin Luther King were not poor; the college students who have been protesting and sitting-in on campus are well-to-do, and even the participants in the ghetto uprisings of the last few years—although hardly affluent—were not drawn from the poorest sectors of the ghetto.

The social protest of the nineteen-sixties has to do with *inequality*, with the pervasive inequities remaining in American life. So far the demand for greater equality has come largely from the young and the black, but I wish to suggest that in the years to come, America will face a demand for more equality in various aspects of life from many other types of citizens—a demand so

From the *New York Times Magazine,* November 3, 1968, copyright © 1968 by The New York Times Company.

pervasive that it might well be described as the "equality revolution."

This demand will take many forms. Some will ask for *equality*, pure and simple; others will press for more *democracy*, for greater participation in and responsiveness by their places of work and their governments; yet others will ask for more *autonomy*, for the freedom to be what they want to be and to choose how they will live. All these demands add up to a desire for greater control over one's life, requiring the reduction of the many inequities—economic, political and social—that now prevent people from determining how they will spend their short time on this earth.

Ever since the Declaration of Independence decreed that all men are created equal, Americans have generally believed that they were or could be equal. Of course, the Constitution argued by omission that slaves were unequal, and we all know that many other inequities exist in America. Undoubtedly, the most serious of these is economic.

About a fifth of the country lives on incomes below the so-called Federal "poverty line" of $3,300 for an urban family of four, and the proportion is higher if the population not counted by the last census (14 per cent of all Negro males, for example) is included. An additional 7 per cent of households, earning between $3,300 and $4,300 a year, are considered "near poor" by the Social Security Administration. Altogether, then, probably about a third of the country is living at or below the barest subsistence level—and about two-thirds of this population is white.

Moreover, despite the conventional description of America as an affluent society, few of its citizens actually enjoy affluence. The Bureau of Labor Statistics estimates that an urban family of four needs $9,376 a year (and more than $10,000 in New York City) for a "modest but adequate standard of living"; but in 1966, 69 per cent of American families with two children were earning less than $10,000. (Their median income was $7,945, although a recent City University study showed the median income of New York families in 1966 to be only $6,684—and those of Negroes and Puerto Ricans to be $4,754 and $3,949 respectively.)

Even $9,400 is hardly a comfortable income, and it is fair to say that today the affluent society includes only the 9 per cent of Americans who earn more than $15,000 a year. Everyone else still worries about how to make ends meet, particularly since the standards of the good life have shot up tremendously in the last two decades.

Of course, income levels have also risen in the last 20 years, and an income of $9,400 would classify anyone as rich in most countries. Even the earnings of America's poor would constitute affluence in a country like India. But comparisons with the past and with other countries are irrelevant; people do not live and spend in the past or in other countries, and what they earn must be evaluated in terms of the needs and wants identified as desirable by the mass media and the rest of American culture. Undoubtedly an advertising man or a college professor who earns $15,000 is in the richest 1 per cent of all the people who ever lived, but this fact does not pay his mortgage or send his children to school. And if *he* has economic problems, they are a thousand times greater for the poor, who have much the same wants and hopes, but must make do with $3,000 a year or less.

The extent of economic inequality is also indicated by the fact that the richest 5 per cent of Americans earn 20 per cent of the nation's income; but the bottom 20 per cent earn only 5 per cent of the income. Although this distribution has improved immeasurably since America's beginnings,* it has not changed significantly since the nineteen-thirties. In other words, the degree of economic inequality has not been affected by the over-all increase in incomes or in gross national product during and after World War II.

There are other kinds of economic inequality in America as

---

* For example, University of Michigan historian Sam Warner reports in "The Private City" that among the minority of Philadelphians affluent enough to pay taxes in 1774, 10 per cent owned 89 per cent of the taxable property.

well. For example, most good jobs today require at least a bachelor's degree, but many families still cannot afford to send their children to college, even if they are not poor. Job security is also distributed unequally. College professors have tenure and are assured of life-time jobs; professionals and white-collar workers earn salaries and are rarely laid off, even in depressions; factory workers, service workers and migrant farm laborers are still paid by the hour, and those not unionized can be laid off at a moment's notice.

Economic inequality goes far beyond income and job security, however. Some executives and white-collar workers have a say in how their work is to be done, but most workers can be fired for talking back to the boss (and are then ineligible for unemployment compensation). Generally speaking, most work places, whether they are offices or factories, are run on an autocratic basis; the employee is inherently unequal and has no more right to determine his work, working conditions or the policy of his work place than the enlisted man in the Army. He is only a cog in a large machine, and he has about as much influence in deciding what he will do as a cog in a machine. Our schools are similarly autocratic; neither in college nor in elementary and high school do students have any significant rights in the classroom; they are unequal citizens who must obey the teacher if they are to graduate.

The poor suffer most from these inequalities, of course. They hold the least secure jobs; they are least often union members; if they are on welfare, they can be made penniless by displeasing the social workers in charge of their cases. And being poor, they pay more for everything. It is well known that they pay more for food (sometimes even at supermarkets) and for furniture and other consumer goods; they also pay more for hospital care, as a recent study in New Haven indicates. They even pay more when they gamble. Affluent Americans can gamble in the stock market, where it is difficult to lose a lot of money except in the wildest speculation. The poor can afford only to play the numbers, where the chance of a "hit" is about 1 in 600, and if they prefer not to participate in an illegal activity, they can play the New York State Lottery, where the chance of winning is only about 1 in 4,000.

Political inequality is rampant, too. Although the Supreme Court's one-man, one-vote decision will eventually result in voting equality, the individuals who contribute to a candidate's election campaign will have far more political influence than others.

Ordinary citizens have few rights in actual practice; how many can afford to argue with policemen, or hire good lawyers to argue their cases, or make their voices heard when talking to their elected representatives? This year's political conventions have indicated once again that the rank-and-file delegates (including those named by political bosses or rigged state conventions) have little say in the choosing of Presidential candidates or platforms. Even the person who is included in a sample of the now-so-important public-opinion polls cannot state his opinion if the pollster's questions are loaded or incorrectly worded.

Finally, there are many kinds of inequality, autocracy and lack of autonomy of which most Americans are not even aware. In many cities, for instance, high-speed mass-transit lines rarely serve poorer neighborhoods and really good doctors and lawyers are available only to the wealthy. Rich or poor, not many people have a say in the choice of TV programs they are shown or in the rates they are charged by electric companies; and who can escape from the poisons in the air?

In a large and complex society, inequality and the lack of control over one's life are pervasive and are often thought to be inevitable byproducts of modernity and affluence. We are learning, however, that they are not inevitable—that there can be more equality, democracy and autonomy if enough people want them.

In the past, when most people earned just enough to "get by," they were interested mainly in higher incomes and did not concern themselves with equality or autonomy in their everyday lives. For example, the poor took—and still will take—any jobs they could get because they needed the money to pay for the week's food and the month's rent. Working-class and lower-middle-class people were, and are, only slightly more able to choose; they take whatever job will provide the most comfortable lives for themselves and their families. But in the upper-middle class, the job is expected to offer personal satisfactions, and upper-middle-class people gravitate to the jobs and careers that provide more equality and autonomy. The huge increase in graduate-school enrollments suggests that many college students want the personal freedom available in an academic career; their decreasing interest in business careers indicates that they may be rejecting the autocracy and lack of autonomy found in many large corporations.

Today, as more people approach the kind of economic security already found in the affluent upper-middle class, they are beginning to think about the noneconomic satisfactions of the job and of the rest of life; as a result, aspirations for more equality, democracy and autonomy are rising all over America.

Some manifestations of "the equality revolution" are making headlines today, particularly among students and blacks. Whatever the proximate causes of college protests and uprisings, the students who participate in them agree on two demands: the right to be treated as adults—and therefore as equals—and the right to participate in the governing of their schools. Though the mass media have paid most attention to the more radical advocates of these demands, equality and democracy are sought not just by the Students for a Democratic Society but by an ever-increasing number of liberal and even conservative students as well.

Similar demands for equality and democracy are being voiced by the young people of the ghetto. Only a few years ago, they seemed to want integration, the right to become part of the white community. Today, recognizing that white America offered integration to only a token few and required with it assimilation into the white majority, the young blacks are asking for equality instead. When they say that black is beautiful, they are really saying that black is equal to white; when the ghetto demands control of its institutions, it asks for the right to have the same control that many white neighborhoods have long had.

And although the call for "participatory democracy" is voiced mainly by young people of affluent origins in the New Left, a parallel demand is manifesting itself among the young blue-collar supporters of Governor Wallace. What they are saying, in effect, is that they are tired of being represented by middle-class politicians; they want a President who will allow the working class to participate in the running of the Federal Government and will get rid of the upper-middle-class professionals who have long dominated the formulation of public policies, the people whom the Governor calls "pseudo-intellectuals."

Many other instances of the equality revolution are less visible, and some have not made the headlines. For example, in the last two generations, wives have achieved near equality in the family, at least in the middle class; they now divide the housework with their husbands and share the decision-making about family expenditures and other activities. Today, this revolution is being extended to the sexual relationship. Gone is the day when women were passive vessels for men's sexual demands; they are achieving the right to enjoy sexual intercourse.

Children have also obtained greater equality and democracy. In many American families, adolescents are now free from adult interference in their leisure-time activities and their sexual explorations, and even preteens are asking to be allowed their own "youth culture."

Man's relationship to God and the church is moving toward greater equality, too. The minister is no longer a theological father; in many synagogues and Protestant churches, he has become the servant of his congregation, and the unwillingness of many Catholics to abide by the Pope's dictates on birth control hides other, less publicized, instances of the rejection of dogma that is handed down from on high. The real meaning of the "God is Dead" movement, I believe, is that the old conception of God as the infallible autocrat has been rejected.

In the years to come, the demand for more equality, democracy and autonomy is likely to spread to many other aspects of life. Already, some high-school students are beginning to demand the same rights for which college students are organizing, and recipients of public welfare are joining together to put an end to the autocratic fashion in which their payments are given to them. Public employees are striking for better working conditions as well as for higher wages; teachers are demanding more freedom in the classroom and—in New York—the right to teach where they choose; social workers want more autonomy in aiding their clients, and policemen seek the right to do their jobs as they see fit, immune from what they call "political interference." The right of the individual to determine his job is the hallmark of the professional, and eventually many workers will seek the privileges of professionalism whether or not they are professional in terms of skills.

Eventually, the equality revolution may also come to the large corporations and government agencies in which more and more people are working. One can foresee the day when blue-collar and white-collar workers demand a share of the profits and some voice in the running of the corporations.

Similar changes can be expected in the local community. Although the exodus to suburbia took place primarily because people sought better homes and neighborhoods, they also wanted the ability to obtain greater control over governmental institutions. In the last 20 years, the new suburbanites have overthrown many of the rural political machines that used to run the suburbs, establishing governments that were responsive to their demands for low taxes and the exclusion of poorer newcomers. In the future, this transformation may spread to the cities as well, with decentralized political institutions that respond to the wants of the neighborhood replacing the highly centralized urban machines. New York's current struggle over school decentralization is only a harbinger of things to come.

Consumer behavior will also undergo change. The ever-increasing diversity of consumer goods represents a demand for more cultural democracy on the part of purchasers, and the day may come when some people will establish consumer unions and cooperatives to provide themselves with goods and services not offered by large manufacturers. Television viewers may unite to demand different and perhaps even better TV programs and to support the creation of UHF channels that produce the types of quality and minority programming the big networks cannot offer.

It is even possible that a form of "hippie" culture will become more popular in the future. Although the Haight-Ashbury and East Village hippies have degenerated into an often-suicidal drug culture, there are positive themes in hippiedom that may become more acceptable if the work-week shrinks and affluence becomes more universal; for example: the rejection of the rat race, the belief in self-expression as the main purpose of life, the desire for a more communal form of living and even the idea of drug use as a way to self-understanding. In any case, there is no reason to doubt that many people will want to take advantage of a "square" form of the leisurely hippie existence—now available only to old people and called retirement—while they are still young or middle-aged. This day is far off, and by then marijuana is likely to have achieved equality with liquor as America's major elixir for temporary escape from reality and inhibition.

These observations suggest that the future will bring many kinds of change to America, producing new ideas that question beliefs and values thought to be sacrosanct. Who, for example, imagined a few years ago that the ghetto would reject the traditional goal of integration or that college students would rise up against their faculties and administrations to demand equal rights? Thus, nobody should be surprised if in the next few years adolescents organize for more freedom in their high schools or journalists decide that their editors have too much power over their work.

These demands for change will, of course, be fought bitterly; protests will be met by backlash and new ideas will be resisted by old ideologies.

Today many argue that college students are still children and should not be given a voice in college administration, just as many say that women do not really need orgasms or that men who help their wives at home are becoming effeminate. Undoubtedly, the defenders of outmoded traditions will argue sincerely and with some facts and logic on their side, but processes of social change have little to do with sincerity, facts or logic.

When people become dissatisfied with what they have and demand something better they cannot be deterred by facts or logic, and the repression of new ideas and new modes of behavior is effective only in the very short run.

But perhaps the most intense struggle between new ideas and old ideologies will take place over America's political philosophy, for a fundamental change is taking place in the values which guide us as a nation. In a little-noticed portion of the "Moynihan Report," Daniel P. Moynihan pointed out that the civil rights struggle, which had previously emphasized the achievement of liberty, particularly political liberty from Jim Crow laws, would soon shift to the attainment of equality, which would allow the "distribution of achievements among Negroes roughly comparable to that of whites."

Moynihan's prediction was uncannily accurate with respect to the civil rights struggle, and I would argue, as he does, that it will soon extend to many other struggles as well and that the traditional belief in liberty will be complemented and challenged by a newly widespread belief in the desirability of equality.

Since America became a nation, the country has been run on the assumption that the greatest value of all is liberty, which gives people the freedom to "do their own thing," particularly to make money, regardless of how much this freedom deprives others of the same liberty or of a decent standard of living. Whether liberty meant the freedom to squander the country's natural resources or just to go into business for oneself without doing harm to anyone else, it was the guiding value of our society.

Today, however, the demand for liberty is often, but not always, the battle cry of the "haves," justifying their right to keep their wealth or position and to get more. Whether liberty is demanded by a Southern advocate of states' rights to keep Negroes in their place or by a property owner who wants to sell his house to any white willing to buy it, liberty has become the ideology of the more fortunate. In the years to come, the "have-nots," whether they lack money or freedom, will demand increasingly the reduction of this form of liberty. Those who ask for more equality are not opposed to liberty *per se,* of course;

what they want is sufficient equality so that they, too, can enjoy the liberty now virtually monopolized by the "haves."

The debate over liberty vs. equality is in full swing, and one illuminating example is the current argument about the negative income tax and other forms of guaranteed annual incomes for the underpaid and the poor. The advocates of guaranteed annual incomes want greater equality of income in American society; the opponents fear that the liberty to earn as much as possible will be abrogated. However, neither side frames its case in terms of equality or liberty. The advocates of a guaranteed annual income rely on moral argument, appealing to their fellow Americans to do away with the immorality of poverty. The opponents charge that a guaranteed annual income will sap the incentive to work, although all the evidence now available suggests that professors and other professionals who have long had virtually guaranteed annual incomes have not lost their incentive to work, that what saps incentive is not income but the lack of it.

Being poor makes people apathetic and depressed; a guaranteed income would provide some emotional as well as economic security, raise hopes, increase self-respect and reduce feelings of being left out, thus encouraging poor people to look for decent jobs, improve family living conditions and urge their children to work harder in school. A guaranteed annual income may reduce the incentive to take a dirty and underpaid job, however, and at the bottom of the debate is the fear of those who now have the liberty to avoid taking such jobs that less-fortunate Americans may be given the same liberty.

In the years to come, many other arguments against equality will develop. We have long heard that those who want more equality are radicals or outside agitators, seeking to stir up people thought to be happy with the way things are. This is clearly nonsensical, for even if radicals sometimes lead the drive for more equality, they can succeed only because those who follow them are dissatisfied with the status quo.

Another argument is that the demand for more equality will turn America into a society like Sweden, which is thought to be conformist, boring and suicidal, or even into a gray and regimented society like Russia. But these arguments are nonsensical, too, for there is no evidence that Swedes suffer more from ennui than anyone else, and the suicide-rate—high in all Scandinavian countries save Norway—was lower in Sweden at last counting than in traditionalist Austria or Communist Hungary and only slightly higher than the rate in *laissez-faire* West Germany or pastoral Switzerland. And current events in the Communist countries provide considerable evidence that the great economic equality which some of these countries have achieved does not eliminate the popular desire for freedom and democracy.

But perhaps the most frequently heard argument is that the unequal must do something to earn greater equality. This line of reasoning is taken by those who have had the liberty to achieve their demands and assumes that the same liberty is available to everyone else. This assumption does not hold up, however, for the major problem of the unequal is precisely that they are not allowed to earn equality—that the barriers of racial discrimination, the inability to obtain a good education, the unavailability of good jobs or the power of college presidents and faculties make it impossible for them to be equal. Those who argue for earning equality are really saying that they want to award it to the deserving, like charity. But recent events in the ghettos and on the campuses have shown convincingly that no one awards equality voluntarily; it has to be wrested from the "more equal" by political pressure and even by force.

Many of the changes that make up the equality revolution will not take place for a generation or more, and how many of them ever take place depends on at least three factors: the extent to which the American economy is affluent enough to permit more equality; the extent to which America's political institutions are able to respond to the demands of the unequal, and—perhaps most important—the extent to which working-class and lower-middle-class Americans want more equality, democracy and autonomy in the future.

If the economy is healthy in the years to come, it will be able to "afford" more economic equality while absorbing the costs of such changes as the democratization of the work place, increased professionalism and more worker autonomy. If automation and the currently rising centralization of American industry result in the disappearance of jobs, however, greater equality will become impossible and people will fight each other for the remaining jobs. This could result in a bitter conflict between the "haves" and the "have-nots" that might even lead to a revolution, bringing about formal equality by governmental edict in a way not altogether different from the Socialist and Communist revolutions of the 20th century. But that conflict between the "haves" and the "have-nots" could also lead to a right-wing revolution in which the "haves," supported by conservatives among the "have-nots," would establish a quasi-totalitarian government that would use force to maintain the existing inequalities.

Although the likelihood of either a left-wing or a right-wing revolution is probably small, even a gradual transformation toward greater equality is not likely to be tranquil. More equality for some means a reduction in privilege for others, and more democracy and autonomy for some means a loss of power for others. Those who have the privilege and the power will not give them up without a struggle and will fight the demand for more equality with all the economic and political resources they can muster. Even today, such demands by only a small part of the black and young population have resulted in a massive backlash appeal for law and order by a large part of the white and older population.

Moreover, whenever important national decisions must be made, American politics has generally been guided by majority rule or majority public opinion, and this has often meant the tyranny of the majority over the minority. As long as the unequal are a minority, the structure of American politics can easily be used to frustrate their demands for change. The inability of the Federal Government to satisfy the demands of the Negro population for greater equality is perhaps the best example. In the future, the political structure must be altered to allow the Government to become more responsive to minority demands, particularly as the pressure for equality grows.

Whether or not such governmental responsiveness will be politically feasible depends in large part on how working-class and lower-middle-class Americans feel about the equality revolution.

They are the ruling majority in America, and if they want more equality, democracy and autonomy, these will be achieved—and through peaceful political methods. If the two classes remain primarily interested in obtaining more affluence, however, they will be able to suppress demands for equality by minorities, especially those demands which reduce their own powers and privileges. No one can tell now how these two classes will feel in the future, but there is no doubt that their preferences will determine the outcome of the equality revolution.

Still, whatever happens in the years and decades to come, the equality revolution is under way, and however slowly it proceeds and however bitter the struggle between its supporters and opponents, it will continue. It may succeed, but it could also fail, leaving in its wake a level of social and political conflict unlike any America has ever known.

What I have written so far I have written as a sociologist, trying to predict what will occur in coming generations. But as a citizen, I believe that what will happen ought to happen, that the emerging demand for more equality, democracy and autonomy is desirable. Too many Americans, even among the nonpoor, still lead lives of quiet desperation, and the good life today is the monopoly of only a happy few. I think that the time has come when unbridled liberty as we have defined it traditionally can no longer be America's guiding value, especially if the right to liberty deprives others of a similar liberty. But I believe also that there is no inherent conflict between liberty and equality; that the society we must create should provide enough equality to permit everyone the liberty to control his own life without creating inequality for others, and that this, when it comes, will be the Great Society.

What Kind of Nation Are We?

by Andrew Hacker

TWO IMAGES WILL live on in our minds. One is of a deranged assassin, coolly aiming a rifle from a Dallas window; the other is of the dignified bearing of the President's wife in the aftermath of her husband's death. Yet before these images take too firm a hold it would be well to recall that neither is representative of American life at this juncture in our history. In a nation of almost 200 million people, there are few minds twisted to the point of senseless violence. And, no less significant, there are few among us with the background and breeding epitomized by a woman like Jacqueline Kennedy. This is an appropriate time, therefore, to ask just what kind of a nation we are.

America is a complex of individuals and institutions, each shaping the values and behavior of the other. While only a quarter of a century from our bicentennial, we are still a new nation, pioneering on many frontiers and haunted by a maze of unresolved problems. Perhaps what characterizes America best is the sense of movement, our readiness to embark on new ventures and our willingness to adjust ourselves to major transformations

From the *New York Times Magazine,* December 8, 1963, copyright © 1963 by The New York Times Company.

in our national life. A description of the country we are, therefore, cannot help but be, in part, a description of the society we are becoming.

We are, first and foremost, a democracy. Indeed, of all the major countries of the world, we are probably the most democratic in feeling and actions. To be sure, the will of the majority is not reflected in many of our political institutions. In the Congress, in state legislatures and in a thousand city halls and county courthouses, well-entrenched minorities have their way. But this is partly because there is no coherent and continuing majority; and even if there were, it is not clear that this aggregate would have a "will" demanding to be translated into legislation.

Perhaps this is only another way of saying that America is not really a "political" nation. For most of us, affairs of state come well down in our list of personal concerns. Only a few petition, picket or even participate on a sustained basis. Democracy, therefore, must be interpreted as a social rather than a political phenomenon. It has meant that a greater and greater number of Americans have been able to ascend to a level of material security and a consciousness of personal worth that in other countries remain the privilege of exclusive minorities.

It is this democratic psychology that troubles critics of America more than anything else. The typical citizen is seen as overly opinionated and overly concerned with what society owes to him. Thus American entertainment is labeled "mass culture." American education is "life-adjustment," and American politics are a series of "popularity contests." Moreover, Americans are seen as demanding as a right admission to higher education, to executive and professional occupations—in fact, to whatever citadels of privilege and prestige society has to offer. The principle of equality is no longer a textbook rubric but the notion that all opinions and talents are of equal quality.

More serious is the charge that there exists little public enthusiasm for those who defy accepted conventions. Our reluctance to tolerate unpopular ideas is a national characteristic having deep roots. Traditional assumptions about innocence until guilt is proven, about the status of those who refuse to incriminate themselves, about the free market-place of ideas—these defenses of individual freedom find little popular support when put to the acid test. While an American majority may be hard to discover on most issues, there is a consensus here. There is the feeling that ideas can be corrupting and that accused enemies of society should be locked up with a minimum of formality. Most Americans share this tough-minded sentiment, and while sincerely shocked by open violence, they are prepared to assume that liberty is for those who are not suspected of abusing it.

Yet critics of the American temper are, in reality, decrying the fact that social classes play so small a role in our national life. Political democracy has a long history in Western Europe but it has been hedged in by class barriers that kept the bulk of its citizens in what was to be their proper stations. Here and there, the feudal tradition lingers on, and the public remains deferential to its "betters."

Those who emigrated to America broke with this tradition of deference, and in so doing embraced the view that everyone has the right to whatever ambitions he cares to set for himself. The foundation stone of American democracy, then, is its classlessness, its opportunities for mobility for tens of millions rather than for a handful in each generation.

Only in America is it so certain that a majority will go on to college, wear a white collar at work and experience a style of life hitherto reserved for those born to privilege. Without a doubt, Michigan State University will never be an Oxford; the executives of United States Steel will not have the polish of the governors of the Bank of England; and the output of Hollywood will fail to rival the standards of a Fellini or a Bergman.

But style, tone and sensitivity are no more typical of a majority of Europeans than of Americans. The difference is that we refuse to keep down whole classes just because their rise may rend the fabric of culture and civility. America pays a price for its egalitarian principles, but the cost is small compared with the vistas and opportunities that have opened to countless citizens.

And, so long as one is a "good American," vague as that criterion may be, religious and ethnic considerations are viewed as secondary. The melting pot is doing its work—especially on those under the age of 40—and while its product may be blandly homogeneous, it must not be forgotten that graduation to the status of American is a marked advance for those who have succeeded in this climb.

Sociologists are fond of talking about the alienation and the quest for "identity" that plague the new middle class and even the affluent working class. Yet for most, the relief from crowded slums, desolate small towns and immigrant identities has been a liberating experience. The problem is how to use this freedom, and not to question its existence.

But talk of a homogeneous national character must be tempered by a recognition that there are still several Americas, certainly so far as values and outlook are concerned.

There is, for example, the America of the provinces, belonging to those who remain on farms and in small towns and medium-sized cities. This continues to be conservative country, adhering to the principles of limited government, small business and the ingrown community. Provincial America is well represented in Congress by both political parties and has a major say in the affairs of our Government. This nation within a nation is on the decline in social and economic terms, but its power in legislative councils gives its viewpoint not only a hearing but often a political veto.

Yet alongside it is a growing metropolitan America, typified by the new suburbs more than the cities and by national corporations as much as by the Executive branch of the Federal Government. Tens of millions of Americans have left small towns and the countryside in recent decades, leaving behind their provincial values and adjusting to the unspoken rules of metropolitan life. The suburbs and the corporations, both destined to grow, require a new style—best characterized as moderate, modern and uncontroversial.

There is not a little soul-searching about the new thinking in economics, education and religion that metropolitan life encourages, and there are serious reactions—"return to fundamentals"—from time to time. Many of the tensions of the nation arise because of the conflicts between provincial and metropolitan values. This confrontation is often expressed in our much-publicized quests for "national values" and our yearning to redis-cover "democratic principles." Yet when all is said and done, the answer is that such goals and principles will be influenced by our new technology and the ways we organize society to meet the needs of the machines we have created.

For looming over the social scene is technological America, a new association of machines and ideas that staggers the imagi-nation. The nation's economy, again more than that of any other country, has transformed patterns of production and employment beyond recognition.

This is testimony to America's willingness to experiment, to take risks and to adapt itself to new organizational forms and productive processes. Neither automation nor the corporations have social theories to justify their existence; they arise from the drawing boards of engineers and the deliberations of executives. The new economy has no ideology, but millions are taking their places in it and partaking of the incomes and advantages it is able to bestow.

The electronic computer is a monument to America's industrial rationality, to the ability of man to apply his reason to the raw materials of the world around him. The thousands of engineers at I.B.M., at A.T. & T. and at N.A.S.A. are not spectacular men, but they have helped to create and utilize technological knowledge on a spectacular scale.

What is clear is that neither they nor others have the knowl-edge or skills to facilitate the adjustment of society to the new machines. A continuing dilemma of history has been the lag between the productive instruments brought into being by hu-man rationality and the failure of that rationality to organize means of coping with the social consequences of industrial prog-ress. Yet another America, a society of losers, has become the discard heap of the new technology.

The affluence of America is widespread but it is far from universal. Millions have failed to find places for themselves in the new economy. Their problem is more than political. The measures devised by liberals—retraining, fiscal reform, expanded welfare services—will have only a marginal impact on the disruptions we are experiencing.

Those of the conservatives—a freer market mechanism, annual balanced budgets, state rather than Federal action—hold out as little promise. Neither public opinion nor our political institutions can fully understand or devise workable solutions for the plight of those left behind in the wake of the new machines.

Much the same can be said of the overriding domestic issue—civil rights. Insensitivity to the liberties of dissenting or deviating individuals is a small problem compared with the unwillingness to accept Negro Americans as full citizens. The irony, it has often been remarked, is that we have not a "Negro problem" but a problem in the anxieties of white citizens. All that has been said about equality, opportunity and assimilation must be construed as applying to white Americans.

Moreover, the white population is patently unequipped, socially and psychologically, to accept their Negro fellow-citizens as if the color of one's skin made no difference. The reasons why the 90 per cent are so fearful of the 10 per cent are complex, but whether we think of bombings in Birmingham or of flights to the suburbs in the North, the root cause is the same: white Americans, most of them newly arrived in status, sense their vulnerability to those whose proximity might endanger their position in society.

Legislation, if and when passed, may create as many problems as it solves. Backed by the authority of Government, Negroes will have access to places formerly closed to them—access, indeed, to the facilities and opportunities of white America. But white Americans will grow increasingly resentful at such intrusions, and tensions will continue to mount. The problem will have to resolve itself, and the changes that take place will be largely outside the realms of law and politics.

Civil rights has laid bare a nation's weakness. The reluctance of the white majority to make personal sacrifices has been revealed, both to themselves and to the world, and it has shown that the traditional critique of democracy—the tendency of the majority to oppress the minority—has yet to be fully answered.

If America is to be summed up, it is a product of democratic and technological forces each acting and reacting to shape the other. Our commitments to a democratic society and to technological innovation are our most marked characteristics, bringing personal freedom and unequaled prosperity but also producing serious tensions and disruptions in our daily lives.

If the events leading to President Kennedy's death were unpredictable and unpreventable, so are many of the larger developments that give form to our national character and institutions. Much of our time and energy is consumed in adjusting to the unforeseen consequences of the ideas we embrace with an enthusiasm that is uniquely American. Our problems are those of success, and our failures are visible because we are continually conscious of the standards we have set and failed to meet.

Suggested Reading

Kenneth E. Boulding, *The Meaning of the Twentieth Century: The Great Transition,* New York, Harper and Row, 1965 (Harper Colophon paperback). The author, an eminent and highly readable economist, is no enemy of technology but would like to have it under sensible direction.

Barry Commoner, *Science and Survival,* New York, Viking, 1966 (Compass paperback). A "scare" book which confuses science and technology (most such books do), but pleads for civic responsibility on the part of scientists—and others.

William A. Faunce, *Problems of an Industrial Society,* New York, McGraw-Hill, 1968 (paperback). A relatively short and mainly nontechnical book. The author, a sociologist, especially emphasizes the problems of automation and of alienation (which he does not equate).

Victor C. Ferkiss, *Technological Man: The Myth and the Reality,* New York, Braziller, 1969. The author, a political scientist, argues that technological man is still subordinate to managerial (bourgeois) man, and thus is not yet master of the machine.

Dennis Gabor, *Inventing the Future,* New York, Knopf, 1964. An eminent English physicist writes in layman's language about the same concerns as those that engage Kenneth Boulding. For Gabor, science excludes social science—which, of course, is wrong.

230 • *Suggested Reading*

Herman Kahn and Anthony J. Wiener, *The Year 2000: A Framework for Speculation on the Next Thirty-three Years,* New York, Macmillan, 1967. This book contains forecasts on just about everything—first on a "surprise-free" basis and then taking account of rather awkward possibilities, such as a thermonuclear war.

Fritz Machlup, *The Production and Distribution of Knowledge in the United States,* Princeton, Princeton University Press, 1962. This book by an economist is somewhat technical at times, but presents valuable information on investments in the "knowledge industry."

Raymond W. Mack, *Transforming America: Patterns of Social Change,* New York, Random House, 1967 (paperback). A serious but charmingly written book by a sociologist about many aspects of social change, including population, race relations, and technology.

Wilbert E. Moore, *The Impact of Industry,* Englewood Cliffs, N.J., Prentice-Hall, 1965 (paperback). This small book discusses the social aspects of modernization in economically (and thus technologically) underdeveloped countries.

Chandler Morse, and others, *Modernization by Design: Social Change in the Twentieth Century,* Ithaca, N.Y., Cornell University Press, 1969. Six authors contribute an essay each on problems of modernization; fairly technical here and there, but mainly clear and a trifle sedate.

Lewis Mumford, *The Pentagon of Power* (*The Myth of the Machine,* Vol. II), New York, Harcourt Brace Jovanovich, 1970. This brilliant book both builds on and supersedes the author's much earlier *Technics and Civilization.* Mumford is not overwhelmed by the marvels of the machine, and persuasively argues that our society should not be.

Alan F. Westin, *Privacy and Freedom,* New York, Atheneum, 1967. A law professor examines such threats to privacy as highly sophisticated electronic "bugs" and the links among data banks which make possible a computer print-out that "tells all" about individuals.

Lynn White, Jr., *Medieval Technology and Social Change,* London, Oxford University Press, 1962. A short but very scholarly book in which the notes almost exceed the length of the text. The text is delightful, however, as it traces the consequences of such inventions as the stirrup for saddled riders, or the three-field system of crop rotation.

Index

A Note on the Editor

Wilbert E. Moore, one of America's best-informed observers of the relation between technology and society, is Professor of Sociology and Law at the University of Denver. Born in Elma, Washington, he studied at Linfield College, the University of Oregon, and Harvard University. Mr. Moore is a past president of the American Sociological Association and a member of the American Philosophical Society and the American Academy of Arts and Sciences, and was until recently a sociologist with the Russell Sage Foundation. His other books include *Industrial Relations and the Social Order; Economy and Society; Conduct of the Corporation; Social Change; Man, Time, and Society;* and *Impact of Industry.*

NEW YORK TIMES BOOKS published by QUADRANGLE BOOKS

AMERICAN FISCAL AND MONETARY POLICY
edited with an Introduction by Harold Wolozin
AMERICAN FOREIGN POLICY SINCE 1945
edited with an Introduction by Robert A. Divine
AMERICAN LABOR SINCE THE NEW DEAL
edited with an Introduction by Melvyn Dubofsky
AMERICAN POLITICS SINCE 1945
edited with an Introduction by Richard M. Dalfiume
AMERICAN SOCIETY SINCE 1945
edited with an Introduction by William L. O'Neill
BLACK PROTEST IN THE SIXTIES
edited with an Introduction by August Meier and Elliott Rudwick
BRITAIN, 1919—1970
edited with an Introduction by John F. Naylor
CITIES IN TROUBLE
edited with an Introducton by Nathan Glazer
THE CONTEMPORARY AMERICAN FAMILY
edited with an Introduction by William J. Goode
THE CORPORATION IN THE AMERICAN ECONOMY
edited with an Introduction by Harry M. Trebing
CRIME AND CRIMINAL JUSTICE
edited with an Introduction by Donald R. Cressey
ECONOMIC DEVELOPMENT AND ECONOMIC GROWTH
edited with an Introducton by James V. Cornehls
EUROPEAN SOCIALISM SINCE WORLD WAR I
edited with an Introduction by Nathanael Greene
THE MEANING OF THE AMERICAN REVOLUTION
edited with an Introduction by Lawrence H. Leder
MODERN AMERICAN CITIES
edited with an Introduction by Ray Ginger
MOLDERS OF MODERN THOUGHT
edited with an Introduction by Ben B. Seligman
NAZIS AND FASCISTS IN EUROPE, 1918—1945
edited with an Introduction by John Weiss
THE NEW DEAL
edited with an Introduction by Carl N. Degler
POP CULTURE IN AMERICA
edited with an Introduction by David Manning White
POVERTY AND WEALTH IN AMERICA
edited with an Introduction by Harold L. Sheppard
PREJUDICE AND RACE RELATIONS
edited with an Introduction by Raymond W. Mack
TECHNOLOGY AND SOCIAL CHANGE
edited with an Introduction by Wilbert E. Moore